EIGHTEENTH CENTURY INVENTIONS

K.T. ROWLAND

EIGHTEENTH CENTURY INVENTIONS

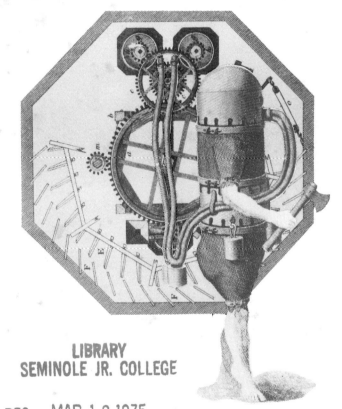

DAVID & CHARLES: NEWTON ABBOT
BARNES & NOBLE BOOKS: NEW YORK
(a division of Harper & Row Publishers, Inc.)

This edition first published in 1974
in Great Britain by
David & Charles (Holdings) Limited
Newton Abbot Devon
in the U.S.A. by
Harper & Row Publishers Inc
Barnes & Noble Import Division

0 7153 6067 1 (*Great Britain*)
06 496015 3 (*United States*)

Printed in Great Britain by
Clarke Doble & Brendon Limited Plymouth

CONTENTS

PREFACE

A study of the list of patents granted during any period in the last three hundred years is indicative primarily of the technology of the age, although for good measure such a list will also contain clues to the prevailing economic and social influences affecting all aspects of contemporary life. The eighteenth century witnessed the beginning of the Industrial Revolution, and many of the historic inventions which were to change fundamentally the structure of society in Europe and North America appear prosaically among the thousand or so patents granted in this period by the British Patent Office. There are some notable exceptions, since some inventors for a variety of reasons failed to take out a patent—an error of judgment that they usually regretted when unscrupulous competitors copied their designs.

This book describes some of the most significant inventions of the century, not all of which were patented, and it also includes many which originated outside the British Isles. The decadence attributed to the *ancien régime* did not affect the inventive capacities of the scientists and engineers of France. The French Army and, in particular, the Corps of Engineers, was a veritable cradle of creative talent that was unmatched elsewhere: and yet all too frequently a brilliant idea remained unexploited in

France through the machinations and procrastinations of the petite bureaucratie. In Britain the climate for development was more favourable since there existed a small yet influential middle class comprising merchants and small manufacturers who usually contrived to sustain and encourage even the most wayward genius to their own exclusive benefit. Two principal groups of inventors emerged in Britain during this period—the millwrights and other skilled tradesmen who formed the core of the inventor-craftsmen, and the mathematical and optical instrument makers who helped to give Britain a position of supremacy in applied science. In temperament and application they differed widely; the former using largely empirical methods with a standard of accuracy that was often ludicrously low, while the latter based their exquisite inventions on acknowledged mathematical principles, and exhibited a degree of skill and craftsmanship that has never been surpassed. And yet by the end of the century, as the Industrial Revolution gained momentum, the gap between these contrasting groups began to narrow, as illustrated by the emergence of the precision machine tool industry, and indications began to appear of the pattern of production which was eventually to provide the basic wealth of the modern industrial state. For geographical and economic reasons these trends soon became discernible in the USA, and early signs of the American aptitude in mass-production techniques can be detected in some of the inventions patented by Americans before 1800.

If British pre-eminence at the end of the century is to be accepted, as it certainly was in steam power, textile engineering and the manufacture of iron, then credit must be given to the enlightened patronage of such bodies as the Society of Arts and the mutual encouragement afforded to those who were Fellows of the Royal Society or members of that select band of savants, the Lunar Society of Birmingham. Nevertheless it was not simply a period of unbroken achievement in an Augustan Age. The triumphs or partial triumphs recorded in these pages obscure the years of fruitless endeavour of many who were less successful, and give no indication of the tragic toll which was to be exacted on those who laboured in the 'dark satanic mills'. Eighteenth-century apathy was eventually replaced by the moral indignation of the latter Victorians, but the scars induced by the enforcement of the factory system on what had previously been a largely agrarian society have remained.

The eighteenth century was a unique period since it signified the beginning of the industrial age, and although progress in most branches of science and technology was steady rather than spectacular there were several fundamental breakthroughs, including Watt's invention of the separate condenser, Arkwright's successful spinning frame and Cort's puddling technique for the bulk production of iron. A curious exception concerned the study of electrical phenomena, with the emphasis on the construction of successively more powerful friction machines— a blind alley which was eventually closed in the final year of the century by Volta and the electric battery. The power and sophistication of some of the later friction machines have been generally overlooked, as has the achievements of most of the eighteenth-century electrical pioneers, who perfected primitive versions of the electric light and the electric telegraph. Similarly it may not be appreciated that the steamship, the submarine and indeed the aeroplane all originated in the eighteenth century, although it was to take a hundred years in the case of the last two before they became a practical reality.

In writing a book of this type one can make no claim to original research, and I can only acknowledge my debt to the numerous authors who have trod this path before and whose more detailed works are listed for further reading. I have also drawn on such recognised sources as the Dictionary of National Biography and the List of Patents and other information supplied by the Patent Office Library. In addition I should like to thank the staff of the Science Museum Library and the Westminster Reference Library for their customary patience and assistance, as well as the following individuals and organisations who have kindly supplied information and illustrations: The Royal Agricultural Society of England, Royal Artillery Institution, The Carron Company, The City Museum and Art Gallery of Birmingham, The Institution of Electrical Engineers, Victoria and Albert Museum, The Radio Times Hulton Picture Library, Musée National des Techniques of Paris and the University of Sheffield. In connection with the above I am particularly grateful to the staff of the Department of Geology at Sheffield University who supplied the photograph of the Spedding Mill on page 150. This was taken on my behalf with the co-operation of the Sheffield City Museum who kindly lent the mill from their collection.

The majority of the illustrations in this book have been reproduced from the volumes of plates issued in conjunction with *The Cyclopaedia* or *Universal Dictionary of Arts, Sciences and Literature* by Abraham Rees FRS, which was published in 1820. This was the first technical encyclopaedia and remains an unsurpassed source of information on the progress of technology up to that time. I wish to record my thanks to David & Charles Ltd for making their volumes available for this purpose, and I should like in conclusion to pay tribute to the artists who were responsible for the original perspective of drawings made during the early years of the nineteenth century. These exhibit a standard of draughtsmanship which has never yet been surpassed.

Brockham,   K. T. ROWLAND
Surrey

Agriculture

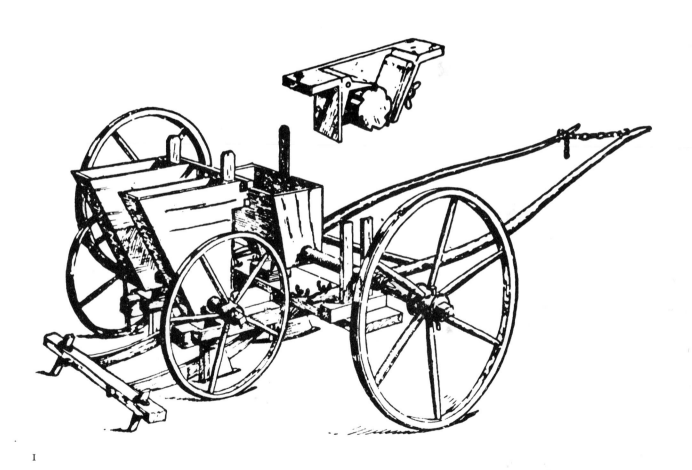

1 SEED DRILL
Jethro Tull (1674–1741)

Until the beginning of the eighteenth century seed was sown by hand as it had been for four thousand years previously and it was not until the problem had been considered by a 'gentleman' farmer that the first tentative steps towards mechanisation were taken.

Jethro Tull, the inventor of the seed drill for which he is justly renowned, was born at Basildon in Berkshire and was educated at St John's College, Oxford. He was called to the Bar in 1699 but did not practise law. Poor health prevented him from following a political career and he took up farming at Howberry, near Wallingford in Oxfordshire. It was at Howberry that he perfected his drill in 1701, although he did not publish an account of his invention until thirty years afterwards. He began work on the machine after his labourers had objected to his plans for sowing the farm with sainform. His basic idea was borrowed from the sound board of an organ and incorporated a groove, a tongue and a spring which permitted a seed to be dropped in the furrow at pre-determined intervals, the action being controlled by a toothed wheel. Tull called the machine a drill simply because hand sowing was known as drilling. He lived to see his drill widely adopted, and other similar machines were built in England and on the continent. There can be no doubt about the advantages gained as a result of Tull's invention since it was estimated that it used only $\frac{3}{4}$ to 1 bushel of seed per acre as opposed to $2\frac{1}{2}$ to $4\frac{1}{2}$ bushels when the seed was hand sown. His ideas on animal husbandry also exerted a great influence on European farming and Voltaire was among the disciples of Tull.

Further Reading
Fussell, G. E., *The Farmer's Tools*, 1952

2 INCUBATION OF EGGS
R. A. F. de Réaumur (1683–1757)

Successful incubation of eggs, an art first practised by the ancient Egyptians, was revived in the eighteenth century, primarily in France. Its general adoption helped to augment the food supplies of the industrial nations and was yet another manifestation of the agricultural revolution of the period.

There were few avenues of scientific endeavour that were not explored by Réaumur during his intensely creative career which spanned the first half of the eighteenth century. Although he is perhaps better remembered for his metallurgical treatise on malleable iron and for his work in many branches of physics, his successful incubator eventually brought immense practical benefits to the expanding populations of France and her other European neighbours. Réaumur followed in the footsteps of the Renaissance inventor Jean Baptiste Porta, who had built a successful incubator at Naples in 1588. Porta's work was based on his knowledge of ancient Egyptian practice, but he abandoned his experiments when he began to acquire the reputation of a sorcerer and attracted the attention of the Inquisition. Réaumur worked

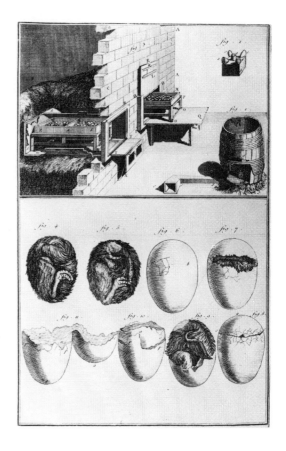

in an age more favouable to scientific innovation, and in fact received the patronage of Louis XV, who took pleasure in assisting the chicks to emerge from their shells. Réaumur's equipment was basically a series of well-ventilated ovens, but it was only after a number of prolonged experiments that he was able to exercise the necessary control of temperature that was essential for successful incubation. To achieve this he invented a thermometer which was one of the most accurate of the day and led to the introduction of the temperature scale which still bears his name. In addition to hatching hens' eggs,

Réaumur experimented with eggs of ducks and various game birds, and even ostrich eggs obtained from the zoo at Versailles. After his death further experiments were carried out by the Abbé Nollet, and later by the Abbé Copineau, who placed eggs in a basket suspended by a wire within a water-jacketed cylinder; an alcohol lamp beneath the eggs provided heat for incubation.

Further Reading
McCloy, S. T., *French Inventions of the Eighteenth Century*, Kentucky, 1952

3 PLOUGH
James Small

James Small was the first ploughwright to attempt to lay down the principles of plough design in a scientific manner, and by using cast iron for the mouldboard he was able to supply farmers all over his native Scotland with identical ploughs that were superior to the traditional Scotch plough.

In 1763 James Small published a pamphlet entitled *Treatise of Ploughs and Wheel Carriages*, in which he attempted to establish certain fundamental principles of plough design within the limits imposed by the tools and manufacturing techniques of the day. Although Small's ploughs are said to have owed much to earlier eighteenth-century models, particularly the Rotherham plough, he was the first ploughwright to give much attention to the design of the mouldboard. In earlier ploughs the furrow slice was turned over only because a very long straight mouldboard was used which was inefficient and made the plough difficult to operate. Individual ploughmen subsequently had curved

mouldboards made to their own design, but Small proposed a universal type of mouldboard that would be generally acceptable and hence could be manufactured in quantity. He made his first mouldboard of wood, and by empirical methods and by calculation he determined the most suitable angle of twist. He also produced a scale showing the required width and curve of the board to cut a certain depth and width of furrow. Once the basic concepts had been established Small's ploughs were made with cast iron mouldboards which was a further improvement over the earlier wooden boards covered with iron sheet. The adoption of the casting process using the same pattern repeatedly resulted in the standardisation that Small was seeking to obtain; his plough became very popular and continued to be used well into the nineteenth century.

Further Reading
Fussell, G. E., *The Farmer's Tools*, 1952

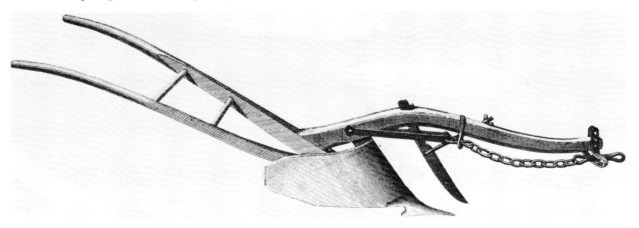

4 DRAINING PLOUGH
Cuthbert Clarke (d 1777)

Clarke's draining plough was one of several introduced during the eighteenth century. Although they were cumbersome and required a number of horses to draw them, draining ploughs constituted the first step in the mechanisation of land drainage.

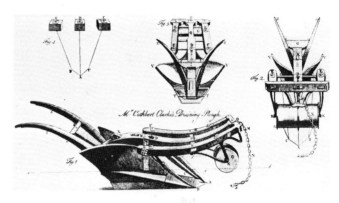

Although certain types of plough were used for drainage work in the seventeenth century, it was normal practice to dig drains by hand. Under the patronage of the Society of Arts, attempts were made in the mid-eighteenth century to introduce draining ploughs for this work. Several prizes were given by the society, one of which was awarded to Cuthbert Clarke of Belford in Northumberland. Clarke's plough was fitted with a roll at the front which prevented the plough from digging too deeply into the earth. It was equipped with three iron coulters or cutters and the frame was of heavy timber construction. Like the other draining ploughs of his contemporaries Knowles and Grey, Clarke's plough was not very effective in heavy clay and was not adopted on a large scale. An account of the plough was included in *The Complete Farmer* published in 1776 and Clarke also described it in a pamphlet entitled *The True Theory and Practice of Husbandry from Philosophical Records and Experience*.

Further Reading
Fussell, G. E., *The Farmer's Tools*, 1952

5 THRESHING MACHINE
Andrew Meikle (1719–1811)

The threshing machine mechanised one of the fundamental processes in the work of the cereal farmer. It was to be rapidly adopted in the early part of the nineteenth century, not however without opposition from farm labourers who considered that it was a threat to their livelihood, particularly during the winter months when the harvest was over.

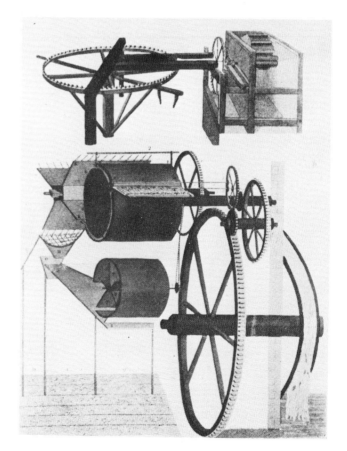

The threshing machine was a composite invention, but it is generally agreed that the chief credit for its perfection in a practical form is due to Andrew Meikle, a Scottish millwright who worked at Houston Mill near Dunbar. Meikle was nearly fifty years old when he obtained his first patent, no896, in association with a Robert Mackell for a machine to dress grain. Ten years later he produced his first threshing machine, but this was similar to an unsuccessful machine patented by a fellow-countryman, Michael Menzies, in 1734. Other relatively unsuccessful machines had been made in the eighteenth century, and one was copied in model form by Francis Kinloch, who farmed at Gilmerton in East Lothian. Kinloch was dissatisfied with his model and he sent it to Meikle. The latter perceived the faults and devised a method of beating the grain to separate it, rather than rubbing it as previous machines had done. To accomplish this he conceived the idea of a rigid revolving drum with fixed scutchers or beaters. He made his own model in which the sheaves were fed on to the drum through two plain feeding rollers, although in later designs these were replaced with fluted rollers. Meikle was then in his seventies, and his ideas were first put into practice on a large scale by his second son George, who made the first threshing machine based on his father's improvements for a Mr Stein, a farmer and distiller at Kilbeggie in Clackmannanshire. The machine worked satisfactorily and orders for further threshers were received from other farmers in the neighbourhood. Andrew Meikle took out patent no1645 in 1788, but for some reason failed to enforce it until it was almost due to expire. In the meantime other millwrights freely copied his machine, and his financial rewards from his invention were small. There can be no doubt that his machine satisfied a long-wanted need, and two years before his death a subscription of £1,500 was raised for him.

Further Reading
Fussel, G. E., *The Farmer's Tools*, 1952
Loudon, J. C., *Encyclopaedia of Agriculture*, 7th edition, 1871

6 TEMPERING IRON PLOUGHSHARES
Robert Ransome (1753-1830)

Ransome's two patents relating to the tempering of iron ploughshares had far-reaching effects on early nineteenth century agriculture. They increased the efficiency of iron ploughs and accelerated the final demise of the wooden ploughshare which still had its supporters even at the end of the nineteenth century.

Despite the metallurgical advances made in iron founding during the eighteenth century, there was a reluctance to use iron for agricultural implements except where it was essential. Iron ploughshares gradually came into service, but they suffered from brittleness, particularly at the cutting edge, and many farmers preferred the traditional wooden plough. The solution was supplied by a Norwich ironfounder, Robert Ransome, in two patents which covered a period of eighteen years. On 18 March 1785 he took out patent no1468 for a method of tempering cast iron ploughshares by wetting the mould with salt water. Four years later Ransome moved to Ipswich and opened the Orwell Works which are still in operation. He carried on the general business of iron founding and may well have manufactured cast iron roofing 'slates' for which he had been awarded a patent in 1783. It is evident however that he still concerned himself very closely with the further improvement of the iron plough, and his work culminated in a second patent, no2736, which was granted to him on 24 September 1803. This involved the chilling of the underside of the ploughshare by casting it in an iron mould, with the rest of the mould being made of sand in the conventional manner. The underside was thus made hard and could be sharpened to retain a cutting edge, while the rest remained fairly ductile and resistant to damage. Ransome's Suffolk iron ploughs became famous; he took out further patents for the improvement of wheel and swing ploughs, and eventually retired from business in 1825. His company, which was carried on by his two sons, later became firmly established as one of the leading manufacturers of agricultural machinery in Britain.

Further Reading
Passmore, J. B., *The English Plough*, Oxford, 1930

7 DOUBLE FURROW PLOUGH
Lord Somerville (1765–1819)

Somerville's double furrow plough invented at the end of the century found favour in districts where the soil was easily worked, and it could increase the area ploughed by a hundred per cent when conditions were favourable.

John Southey, the fifteenth Lord Somerville, was a leading figure in British agriculture from the 1790s through to the end of his life. He was educated at Harrow and St John's College, Cambridge, and became a friend of the younger Pitt. In 1793 he was appointed a member of the Board of Agriculture and three years later, partly due to Pitt's influence, he became president. To Somerville the post was no sinecure for, in addition to being an able administrator, he was a practical farmer with an insight into mechanical problems relating to farm implements. In his early twenties he had invented a single furrow plough, but he is best known for a double furrow plough which he devised in 1798 during the last year of his period of office as president of the board. This was not the first double furrow plough but seems to have been the most successful, and was used in the nineteenth century in certain parts of the country long after Somerville's death. Potentially

it offered a major advantage since it should have enabled a field to be ploughed in half the time, but in practice this was not always the case. It placed a greater strain on the team than a single furrow plough and many authorities considered that it should be limited to light soils or to where a four-horse team was available. Adjustable plates were fitted to the extremities of the mouldboards—a practice introduced by Somerville in his single plough to help turn over the soil—but they were not very effective. Somerville received a Court appointment after relinquishing his post at the Board of Agriculture, and became a close friend of George III who shared a common interest in the development of agriculture. After the king, Somerville became the largest importer and breeder of merino sheep in England, and in 1802 he organised an agricultural show in London which he promoted for many years, mainly at his own expense.

Further Reading
Loudon, J. C., *Encyclopaedia of Agriculture*, 7th edition, 1871

IRON AND STEEL

8 IRON SMELTING
Abraham Darby I (1677-1717)

The use of coke in iron smelting was first accomplished successfully by the Darbys of Coalbrookdale. One of the consequences was that much larger castings, such as the cylinders for the Newcomen atmospheric engines, could be made more easily in iron.

Iron smelting using a mineral fuel such as coke was first attempted in the seventeenth century and several patents were taken out relating to the process. These were only partly successful, and it was not until the Darbys introduced improved blast furnace techniques at Coalbrookdale that the benefits of using coke were fully exploited. Some doubt originally existed regarding the inventor of the Coalbrookdale method, but it is now generally agreed that the credit should be given to Abraham Darby I, who moved from Bristol to Coalbrookdale in 1709 with the intention of setting up as a manufacturer of cast iron pots. At first Darby used charcoal as fuel in the traditional manner, but this was in short supply and he turned to coal as an alternative. This had to be transformed into coke to remove the harmful impurities, the smelting operation also required the presence of a flux in the form of lime or limestone. Both coke and limestone had been used spasmodically in the past and Darby made no attempt to patent the process. The success of his operation was probably due to the use of an effective and continuous furnace blast. It is not known whether the bellows were manufactured to his own design, or if they owed anything to the double-handed bellows invented by Captain Savery in 1705. From contemporary accounts they were substantial and superior in power to those in general use. Abraham Darby I died in 1717, but his sons were too young to carry on the business and it was sold for a paltry sum. Eventually the Darbys re-entered iron founding and the tradition was successfully revived in Coalbrookdale under the leadership of Abraham Darby II. The younger Darby expanded the foundry and started to manufacture large iron castings including cylinders for atmospheric engines which had originally been made in brass. A waterdriven boring machine was installed, and eventually cylinders with diameters of up to 6ft were cast and bored to a reasonable degree of accuracy.

Further Reading

Ashton, T. S., *Iron and Steel in the Industrial Revolution*, Manchester, 1951

Pannell, J. P. M., *An Illustrated History of Civil Engineering*, 1964

9 CRUCIBLE STEEL
Benjamin Huntsman (1704-76)

It is unlikely that the expansion of the machine tool industry in the early years of the nineteenth century, which played such a crucial part in the consolidation of the Industrial Revolution, would have been possible but for the availability of inexpensive and efficient cutting tools formed from crucible steel bar.

One of the most important metallurgical advances in the eighteenth century was the invention of crucible steel by Benjamin Huntsman, originally a clockmaker in Doncaster. Because the blister steel in use at the time was often unsatisfactory for springs and pendulums, Huntsman began to experiment on his own accord to produce steel that would meet his specifications. In 1740 he moved to Handsworth, a few miles south of Sheffield, because fuel was more easily obtainable and continued to carry out his investigations in secret over many years. Huntsman's process was relatively simple since it consisted of melting scrap or blister steel in a crucible and removing as much of the impurities as possible, but he was confronted by two major problems; he required a clay capable of being moulded into a crucible that could withstand intense furnace heat for about five hours, and he needed higher quality scrap steel than could be obtained from local sources. By mixing local earths with clay from the Stourbridge district, he eventually found a suitable crucible material, and he solved the problem of the scrap metal charge by using Swedish iron imported through Hull. One of the critical parts of the smelting operation was the addition of a flux, the

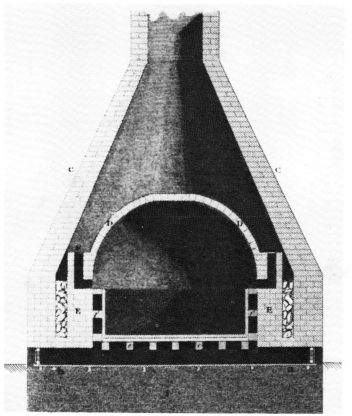

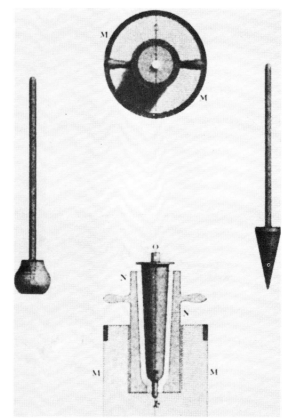

nature of which was a closely guarded secret. Huntsman began to make cast steel successfully in the late 1740s, producing a steel of uniform carbon content that was free of slag, but the local cutlers complained that it was too difficult to work. For a time most of his output was exported to France but, when English merchants began to purchase cutlery from across the channel, the Sheffield cutlers became alarmed and tried unsuccessfully to get the export of crucible steel prohibited. Huntsman received offers to erect a furnace in the Birmingham area but eventually in 1770, when demand increased, he moved his business to Attercliffe on the northern side of Sheffield. He attempted to keep the process secret by excluding strangers from his foundry and carrying out all casting operations at night. One of his rivals, Samuel Walker, is reported to have learnt

the details while disguised as a tramp. Certainly other steel manufacturers were using Huntsman's process before his death, and by 1787 at least seven crucible steel companies were listed in the Sheffield Directory. Nevertheless, the inventor appears to have kept at least some of the secrets to himself as Huntsman steel was more highly regarded than its competitors for many years. The demand came chiefly from manufacturers of razors, pen-knives and other cutlery items requiring a fine edge. It was also used, appropriately, for watch springs and for metal cutting tools.

Further Reading

Ashton, T. S., *Iron and Steel in the Industrial Revolution*, Manchester, 1951

Rolt, L. T. C., *Tools for the Job*, 1965

10 ROLLERS FOR MANUFACTURING IRON BARS
Henry Cort (1740–1800)

Cort's patent for grooved rolls preceded his patent relating to the puddling process by one year, but the two are inseparably linked and together founded British industrial supremacy in iron manufacture for the next century.

In his iron manufacturing experiments carried out at Fontley in Hampshire (ref 11), Henry Cort first introduced and patented the use of grooved rolls for bar manufacture before obtaining a second

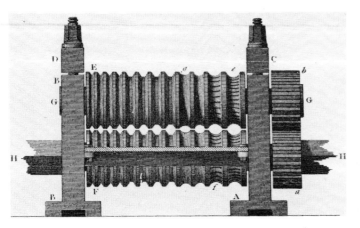

patent to cover the iron puddling process. The apparatus consisted essentially of a pair of rolls placed one above the other and connected at one end by a simple gear train. The lower or driving roll was linked to a source of power such as a water wheel, or in the case of small sets the rolls could be turned by hand. The billets or 'faggots' of iron, after being raised to red heat in a furnace, were passed in succession from one side of the rolls to the other through a series of grooves, each of which was progressively smaller until the required shape was attained. The rolls were made interchangeable; sets with different profiles for round, square, flat and, later, hexagonal bars could be fitted in the same machine. Cort's method of bar manufacture was rapidly adopted by the iron-masters of the day, as also was the puddling process, but for reasons explained on page 21 Cort was unable to enjoy any great material advantage from either of his patents. The originality of his inventions was

challenged; and there is no doubt that he was not the first in Britain to propose the use of grooved rolls for bar manufacture, a patent to this effect having been granted in 1728 to a Mr John Payne. There is no evidence, however, that Payne ever constructed a bar mill, and Cort, whose talent lay in developing and improving the ideas of others, is generally credited with its invention. There can be no dispute about its significance in relation to the output of bar iron in Britain. A tilt hammer with difficulty could forge a ton of bars in twelve hours, while a bar mill could produce up to fifteen tons in the same period, and of superior quality.

Further Reading

Panell, J. P. M., *An Illustrated History of Civil Engineering*, 1964

Ashton, T. S., *Iron in the Industrial Revolution*, Manchester, 1951

11 IRON SMELTING
Henry Cort (1740–1800)

Cort's invention of an efficient and cheap method of making iron helped to shape the course of the Industrial Revolution and exerted a tremendous influence on every branch of technology until the advent of Bessemer steel.

Henry Cort was born in Lancaster but moved to London when a young man. By 1765 he was established as an agent for the Navy in Surrey Street, just off the Strand. At that time the quality of English iron was considered inferior to that of Swedish and Russian iron, and Cort, in the course of his duties, noted that almost all government supplies were imported. When he was thirty-five,

he gave up his business as a naval agent and took over an ironworks at Fontley near Titchfield in Hampshire. He had previously started to investigate improved methods of iron manufacture and there was now an added incentive as the price of continental iron rose sharply. Due probably to his earlier connections with the Navy, Cort was able to obtain large contracts to supply Portsmouth Dockyard with iron hoops for use with masts and spars. He agreed to take back scrap hoops as part payment and began experiments to convert the scrap into iron that could be sold again. Cort's first patent, no1351, dated 17 January, 1783, related to the use of a reverberatory

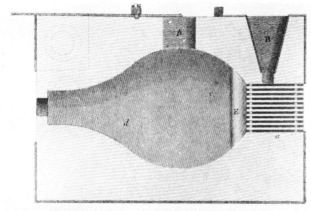

material was pig or cast iron which was again heated in a reverberatory furnace using ordinary pit coal instead of the much scarcer charcoal, the flames being reflected on to the charge from the sloping roof of the furnace. Access doors were provided so that the furnaceman could stir the charge periodically, and oxygen in the air decarbonised the melt to produce malleable iron. The final stage of the process was carried out under a tilt hammer after heating in a separate furnace, and the iron was then rolled into plate or bars. It is generally acknowledged that Cort's methods were not fundamentally new; he took ideas from his predecessors and developed them with his own improvements into a practical iron manufacturing process capable of being applied on a large scale. Within a decade, Britain became the premier iron-producing country in the world and the quality of British wrought iron was universally esteemed. Cort, as many other inventors before and since, received little financial benefit from his patents. His partner Adam Jellicoe died suddenly in 1789 when heavily in debt and Cort's patents, which had previously been given as collateral security, were appropriated by the Crown Offices. In 1794, on the advice of William Pitt, he was granted a pension of £200—a miserly compensation which did no credit to the ironmasters or the government of the day.

Further Reading

Pannell, J. P. M., *An Illustrated History of Civil Engineering*, 1964

Ashton, T. S., *Iron and Steel in the Industrial Revolution*, Manchester, 1951

furnace for heating old mast hoops and other old iron bars to fire welding temperature, the bars being subsequently forged under a tilt hammer and passed through a rolling mill before finally being fed between a series of grooved rollers (ref 10). In the following year Cort obtained a second patent, no1420, dated 13 February, which was even more important since it covered the process of dry puddling—one of the greatest single advances ever made in iron technology. This time the basic

STEAM POWER

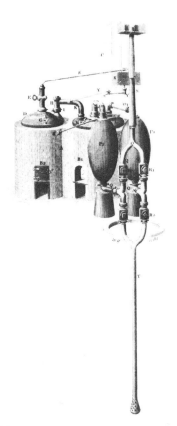

12 STEAM PUMP
Thomas Savery (1650–1715)

Savery's steam pump was the first to perform useful work and to use a separate boiler, but it was unsuited for its intended task of draining mine workings.

Although Savery's steam pump was patented in the closing years of the seventeenth century, it was not until 1701 that the apparatus reached its final form as envisaged by its inventor. Savery was a military engineer with the rank of trench master. He possessed an agile and inventive mind which ranged far beyond the limitations imposed by his profession. In 1696 he invented a machine for polishing plate glass, and in the same year he devised a method of propelling ships by paddles operated from a capstan. His most celebrated invention—the steam pump—was the subject of patent no356 which was granted to him on 25 July 1698. His original model consisted of a single boiler and receiver, but alterations were made and in June 1699 he demonstrated a pump with two receivers to the Royal Society. Further modifications were carried out, and in 1701 Savery wrote a pamphlet entitled *The Miner's Friend* which was published in the following year and included a description of the pump in its final form. His basic unit consisted of a boiler which supplied steam to a closed vessel, the steam being condensed by cold water poured over the outside of the vessel thus creating a vacuum. Water was drawn up through the suction pipe by the vacuum into the vessel, which was fitted with a non-return valve at the bottom. When steam was admitted to the vessel a second time in the cycle, the water was forced out through a second non-return valve and up the discharge pipe. The addition of a second vessel speeded up the operation, since one was under vacuum while the other was discharging to atmosphere. A second and smaller boiler was added which served in effect as a feed heater or reserve feed tank, since it was only brought into service when the test cocks on the main boiler showed that the water level was low. It was Savery's intention, as shown by the title of his pamphlet, to use his pump in mines to prevent flooding, but it proved unsuitable for this purpose. The maximum suction lift was only about 25ft which meant that the receiver and furnace were required to be well below ground. In addition the low steam pressure generated in his primitive boiler restricted the discharge lift and, to be effective in the deeper mines, a series of pumps would have been necessary. Savery set up a workshop at Salisbury Court, off Fleet Street in London, and subsequently installed one of his pumps in York Buildings near the Strand to supply the neighbourhood with water from the Thames. This, however, gave endless trouble, and Savery attempted to improve the lift by increasing the steam pressure, but his design demanded too much of the workmen and materials of his day. In 1705 he appears to have given up the project and devoted himself to other interests. In that year, through the patronage of Prince George of Denmark, he was appointed to the office of treasurer of the Seamen's Hospital, and in 1714, a year before he died, he became surveyor to the waterworks at Hampton Court. Improved versions of his pump were later used in Manchester in conjunction with water wheels for driving machinery in cotton mills, and an apparatus on the same principle was patented by William Blakey in 1766 (ref 16).

Further Reading

Law, R. J., *The Steam Engine*, Science Museum Booklet, 1965

Rolt, L. T. C., *Thomas Newcomen*, Newton Abbot, 1963

Galloway, R., *A History of Coal Mining in Great Britain*, 1882, reprinted Newton Abbot, 1969

13 STEAM PUMP
Denis Papin (1647–c 1712)

At the end of the seventeenth century Papin suggested correctly the course which the future development of the steam engine might follow but, although his own steam pump incorporated a piston, it exhibited too many of the features and defects of Savery's pump (ref 12) and did not lead to a mechanically effective engine.

In 1695 Dennis Papin, who was a professor of

mathematics at Marbourg in Southern Germany, suggested the idea of utilising the expansion and contraction of steam to form a partial vacuum on the underside of a piston moving in a cylinder, the piston being forced down by atmospheric pressure acting on the upper side. The device was intended to solve the perennial problem of pumping out water from deep mine workings and the mode of operation proposed by Papin was, in fact, the one successfully adopted by Newcomen a decade or so later (ref 14). Papin's ideas were made known to Charles, the Landgrave of Hesse, who encouraged him to begin experiments; the work started in 1698 but progress appears to have been slow and it was not until 1705, after Papin had seen an engraving of Savery's pump, that he made any substantial advances. He published a treatise on the subject in 1707 and the pump in its final form obviously owed much to Savery's engine, although it incorporated a piston moving in a cylinder which was an advance on Savery's vacuum chamber. Drawings show that Papin's pump was comparitively small and compact. It consisted of a cylindrical boiler equipped with a safety valve—another invention attributed to Papin. The boiler supplied steam to a cylinder enclosed at the top and containing a floating piston. The base of the cylinder was formed into a curved tube which connected with a second larger cylinder. Water from a reservoir primed the steam cylinder and,

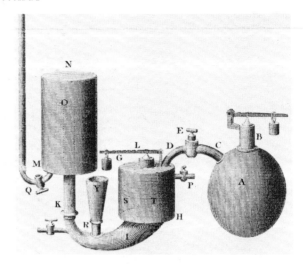

when the boiler stop valve was opened, steam pushed down the piston and forced the water into the second cylinder, which acted as a receiver, and then out through a discharge valve. Various priming and drain cocks were fitted, which had to be manually operated to ensure that the cycle was completed satisfactorily. This was a disadvantage and restricted the application of the pump, although Papin proposed that it might be used to drive a water mill.

Further Reading
Tredgold, T., *Tredgold on the Steam Engine*, 1851

14 ATMOSPHERIC STEAM ENGINE
Thomas Newcomen (1663-1729)

Newcomen's atmospheric steam engine was the first true steam engine in which mechanical work was performed by a piston moving in a cylinder. It was used almost exclusively by the mining industry to pump out water from much greater depths than had previously been possible, and as a result the output of many mines, particularly those in Cornwall, was considerably increased.

Thomas Newcomen was the first of the great craftsmen-inventors of the eighteenth century who were responsible for the development of the steam engine. Newcomen combined the trades of iron-monger and blacksmith in the small Devon port of Dartmouth, and it is thought his first acquaintance with steam came when he was called to advise on the maintenance of one of Savery's steam pumps installed in a mine in the neighbouring county of

Cornwall. Savery's apparatus had obvious limitations, and Newcomen, with the aid of an assistant, John Calley, began a series of experiments that ultimately resulted in the first atmospheric engine which was erected at a colliery near Dudley Castle in Staffordshire in 1712. In this type of engine low pressure steam, usually 2lb/sq in above atmospheric pressure, was admitted to the underside of a piston designed to move up and down in a cylinder that was open at the top to the atmosphere. At a pre-determined point in the cycle, a jet of water was admitted to the cylinder beneath the piston, thus condensing the steam and creating a vacuum; atmospheric pressure acting on top of the piston then forced it down to produce the power stroke. The piston was connected to one end of an overhead beam and the mine pump rod was attached to the

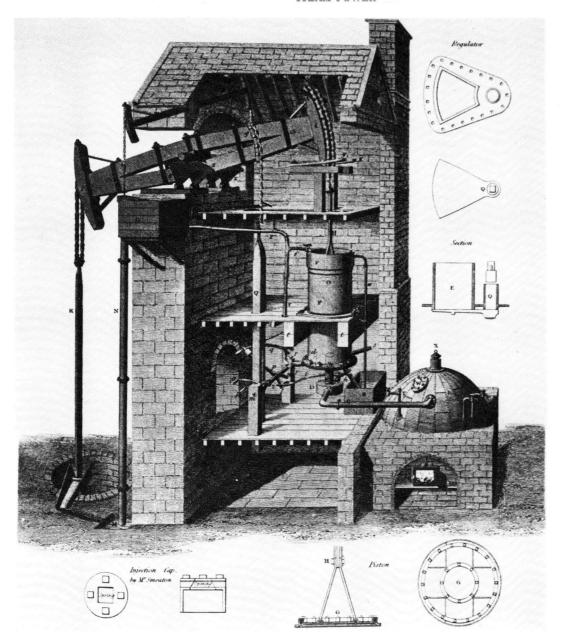

other end. The thermal efficiency of the Newcomen engine was extremely low, but the design was reliable and took into account the limitation of the tools and manufacturing techniques of the day. Difficulties were experienced in boring the cylinders of some of the earlier engines which were cast in brass and probably finished off by hand. After about 1724 cast iron was used for this purpose, and both the Coalbrookdale ironworks and later the Carron ironworks in Scotland became proficient in cylinder manufacture. Newcomen was unable to patent his invention as Savery held a master patent which covered all means of raising water by fire. The two men, however, collaborated during Savery's lifetime, and afterwards Newcomen engines continued to be built, although royalties were paid to those holding the rights of Savery's patent until it expired in 1733.

Further Reading

Law, R. J., *The Steam Engine*, 1963

Barton, D. B., *A History of Copper Mining in Cornwall and Devon*, Truro, 1968

Rolt, L. T. C., *Tools for the Job*, 1965

Ashton, T. S., *Iron and Steel in the Industrial Revolution*, Manchester, 1951

15 STEAM ENGINE
Jacob Leupold (1674–1727)

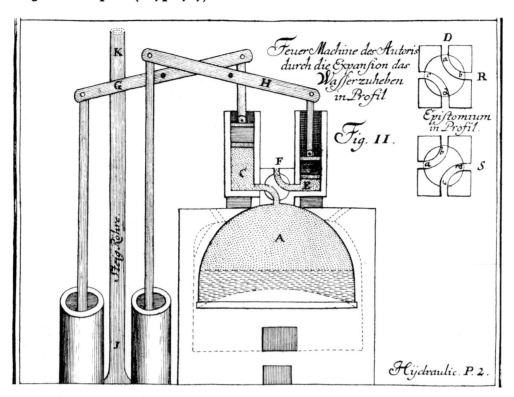

Leupold's engine was the first in which steam was intended to perform mechanical work by moving a piston in a cylinder.

Jacob Leupold, a native of Leipzig, was erroneously credited with the invention of the first high pressure steam engine by a number of early authorities on the history of engineering. Leupold's proposed engine, of which he published details in 1724–7, was not however intended to be driven by steam generated at pressures comparable to those adopted by Trevithick and Oliver Evans in the early years of the nineteenth century. The pressure was only a few pounds above atmospheric, but the design was nevertheless remarkable since the work was to be performed by steam pressure acting on a piston without the benefit of a vacuum. This was a complete contradiction of the policy of Newcomen which was later reinforced by Watt with his invention of the separate condenser (ref 18), and in this respect Leupold's engine was the true antecedent of the small high-pressure, non-condensing engines that were eventually built. His design was remarkably sophisticated for the time, and included a four-way cock to control the admission and emission of steam.

Two cylinders of equal size were arranged above the boiler and the four-way cock was interposed between the boiler and cylinders. Its purpose was to admit steam to the underside of the piston of one cylinder and allow the steam in the other cylinder to exhaust to atmosphere. When each piston was raised it caused a horizontal lever to pivot, which was connected to the plunger rod of a displacement pump. Both pistons were moved alternately, and the intention was to produce a continuous stream of water. There is no record of this type of machine ever entering service, and it is intriguing to speculate on the subsequent development of the steam engine had it done so. Leupold was demanding degrees of accuracy that may well have been beyond the capabilities of the craftsmen and tools of the day. The pistons would have had to be steam-tight and the cylinders bored to much closer tolerances than those allowed for the Newcomen engines—conditions which almost defeated Watt fifty years later.

Further Reading
Tredgold, T., *Tredgold on the Steam Engine*, 1851
Dickinson, H. W., *The History of the Steam Engine*, 2nd edition, 1963

26

16 STEAM ENGINE AND BOILER
William Blakey

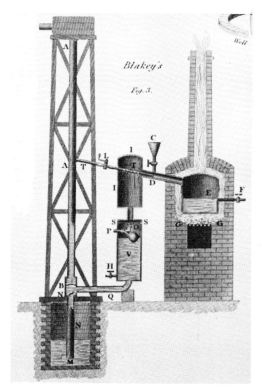

Blakey's steam engine was an improved version of Savery's pump and made little contribution to the general development of steam power; his boiler, patented separately, was however one of the earliest water tube boilers to be built.

William Blakey was granted patent no848 in 1766 for a steam engine which was in effect an improved version of the steam pump designed by Thomas Savery at the end of the seventeenth century (ref 12). The apparatus consisted of a boiler, a cylinder in which the steam was condensed by a jet of water, an air vessel and a suction and discharge pipe. The principal differences between Blakey's engine and that of Savery were the introduction of the air vessel and the use of oil on the surface of the water in the steam vessel or cylinder to reduce the condensation of steam. Watt was aware of Blakey's engine but did not consider that it was a serious competitor for service in the mines. Blakey afterwards patented a water tube boiler in Holland. This was primitive in conception since it contained only three tubes in series, of which only the lower one was full of water. The tubes were made of copper sheet brazed at the seams, and in practice it was found difficult to keep the joints steam tight. The boiler, like the engine, found no great application but it represented a step towards the efficient water-tube boilers of the late nineteenth century.

Further Reading

Robinson, E. and Musson, A. E., *James Watt and the Steam Revolution*, 1969

Dickinson, H. W., *A Short History of the Steam Engine*, 2nd edition, 1963

17 STEAM CARRIAGE
Nicolas Joseph Cugnot (1725–1804)

Cugnot's steam carriage was the first vehicle to be driven by a steam engine and ranks as a major landmark in the history of transport, although lack of financial support at the critical moment due to the vagaries of French politics prevented its subsequent development.

Nicolas Joseph Cugnot, a French military engineer, first began experiments to build a steam-propelled road vehicle while serving at Brussels. In 1763 he was transferred to Paris where eventually his work was brought to the notice of Gribeaval, the inspector general of artillery and the Duc de Choiseul who was minister of war. The minister decided to support Cugnot's experiments with government money and instructed that a full-size vehicle should be built in place of the model on which Cugnot was working. As soon as the larger vehicle had been completed, a demonstration was held before the minister and a large number of spectators. The trials were successful, the vehicle carrying four passengers and attaining a speed of between two and three miles per hour. The authorities were so pleased with the performance that they ordered Cugnot to construct a second and more powerful vehicle capable of transporting a load of four or five tons at a speed of at least two miles per hour, their intention being to use the machine as a tractor to draw heavy artillery into position. Cugnot's second machine, which can still be seen in the Conservatoire National des Arts et Metiérs in

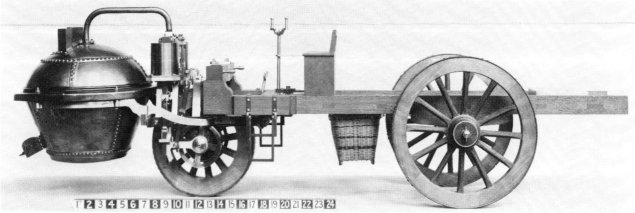

1 2 3 4 5 6 7 8 9 10 11 12 13 14 15 16 17 18 19 20 21 22 23 24

Paris, was completed in July 1771 at a cost of 20,000 livres. It was equipped with three wheels, with the boiler and engine located over the front wheel. The engine consisted of two cylinders with pistons connected to the front or drive wheel through a system of ratchets and levers. The pistons were arranged to operate alternately to give a continuous movement and a fluted iron tyre was fitted around the circumference of the wheel to increase traction. The vehicle had certain inherent weaknesses in design; its tricycle construction made it unstable, and the driver had to stop periodically to add fuel to the furnace and top up the boiler with water. It was nevertheless a remarkable achievement

and it is clear that Cugnot was on the verge of perfecting a major invention which might have resulted in immeasurable benefits to France had further development taken place. Unfortunately, by one of those odd quirks of history, Choiseul, Cugnot's principal backer, fell from power before the second machine was completed and his successor showed no interest in the vehicle. It remained under a shed at the arsenal for thirty years until it was moved to its present resting place in the museum.

Further Reading
McCloy, S. T., *French Inventions of the Eighteenth Century*, Kentucky, 1952

18 SEPARATE CONDENSER AND OTHER IMPROVEMENTS TO THE STEAM ENGINE
James Watt (1736–1819)

Watt's jet condenser is generally recognised as the key invention of the Industrial Revolution. Together with the other improvements included in the historic patent of 1769, it increased the performance of the mine pumping engine developed by Newcomen and led the way to rotative motion— thus providing sustained power to drive the mills of the cotton and woollen industries that were to follow from the inventions of Arkwright, Hargreaves and Crompton.

James Watt's first acquaintance with a steam engine was when he was asked to repair a working model of a Newcomen atmospheric engine (ref 14) while employed as an instrument maker at Glasgow University. Watt not only repaired the engine, but he set out to improve it. In 1765, after he had been working on the problem intermittently for about a year, he conceived the idea of a separate condenser to replace the wasteful practice of

condensing the steam on the underside of the piston by means of a jet of water directed into the cylinder itself. Watt could see that the original Newcomen method was grossly inefficient since the cylinder was successively heated and cooled at every stroke. Translating his ideas into practice was not easy, although he was a skilled craftsman. He received encouragement and financial support first from Dr Joseph Black, who held the Chair of Chemistry at the University, and afterwards from John Roebuck then one of the general managers and principal shareholders of the recently formed Carron Company. Watt patented his proposals in 1769; the patent no 913 was entitled 'A New Method of Lessening the Consumption of Steam and Fuel in Fire Engines' which was a prosaic enough description of an invention that was to set off a train of events which revolutionised the economic, social and commercial

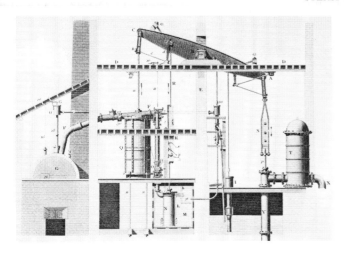

life of Europe. In addition to the condenser, which took the form of a nest of tubes, the patent made provision for an air pump to maintain a vacuum in the condenser, a steam jacket around the cylinder and a stuffing box to prevent steam from escaping to the atmosphere where the piston rod passed through the cylinder cover. The latter was another fundamental innovation of Watt's, for unlike the Newcomen engine, where the top of the piston was open to the atmosphere, his cylinder was enclosed and the driving force to move the piston was steam—albeit low pressure steam. A small engine with an 18in diameter cylinder was built in secrecy in a shed behind Roebuck's house at Kinneil and the resources of the Carron Company were put at

Watt's disposal. Difficulty was experienced in machining the cylinder and Smeaton's boring mill, which had successfully bored a number of Newcomen cylinders, was unequal to the task. In 1773 Roebuck was declared bankrupt and Matthew Boulton, an established Birmingham manufacturer and one of the principal creditors, took over Roebuck's share of the patent. This was to be the turning point in both their lives, with Boulton's industry and business acumen and Watt's mechanical genius making their partnership a formidable combination. The Kinneil engine was finally completed in Birmingham with a cylinder bored on John Wilkinson's original boring machine in April 1775. In the same year the partners shrewdly petitioned parliament for an extension to Watt's original patent and this was granted for a further twenty-five years, until 1800. Production of pumping engines began and orders poured in from the Cornish copper mines and the coal mines of the North when it was seen that the fuel consumed was less than one third of that required to drive a Newcomen engine of similar size.

Further Reading

Law, R. I., *The Steam Engine*, 1965
Rolt, L. T. C., *Tools for the Job*, 1963
Robinson, E. and Musson, A. E., *James Watt and the Steam Revolution*, 1969

19 SUN AND PLANET GEAR FOR STEAM ENGINES
James Watt (1736–1819)

The sun and planet gear was a mechanical expediency devised by Watt to serve in place of the crank which he was debarred from using by patent protection. Although this gear was only fitted to the early Watt rotative engines, it is of historic importance in steam engine development since it was used on all the first generation of mill engines built at Soho.

After the success of the Kinneil engine and the two full-size pumping engines that followed in 1775–6, Boulton, Watt & Co were kept busy dealing with inquiries for pumping engines for the mines of Cornwall and the North. The idea of converting his engine to produce rotary instead of reciprocating motion was never far from Watt's mind; the crank was the obvious solution and, as it

had been in use for at least a hundred years to actuate lathes for wood turning, Watt did not think it necessary to take out a patent. An experimental model incorporating a crank was erected at Soho in 1779, and it was afterwards alleged by Watt that one of his assistants, Richard Cartwright, passed the details to James Pickard of Snow Hill in Birmingham. Pickard proceeded to apply the crank to an atmospheric engine and was granted a patent which covered the use of the crank and counterweight to produce rotary motion with all types of steam engine. Although Watt was bitter in his condemnation of Pickard, he decided not to challenge the patent and set to work to find an alternative solution. The result was his multi-part patent no1306, dated October 1781, which covered no less than five methods of obtaining rotary motion in a steam engine. Several of these were thinly disguised versions of the crank and were never applied in practice; only one in fact, the sun and planet or epicyclic gear, was actually used to any considerable extent. The planet wheel was fixed rigidly to the connecting rod and made to revolve around the perimeter of the sun wheel keyed to the driven shaft. If both wheels were the same size then the driven shaft would make two revolutions for every double stroke of the engine. A single acting rotative engine incorporating the sun and planet gear was erected for John Wilkinson at Bradley Forge to operate a tilt hammer in March 1783 and other ironmasters, such as Hallen of Bewdley, Rathbone of Coalbrookdale and James Spedding of Seaton, placed similar orders. Pickard's patent expired in 1794 and Watt was then able to use the crank, although engines with the sun and planet gear were built as late as 1802 at Soho.

Further Reading

Law, R. J., *The Steam Engine*, 1965
Rolt, L. T. C., *James Watt*, 1962
Dickinson, H. W., *A Short History of the Steam Engine*, 2nd edition, 1963

20 PARALLEL MOTION FOR STEAM ENGINES
James Watt (1736–1819)

The patent granted to Watt in 1784 included his celebrated parallel motion which rendered the engine double acting and enabled it to push as well as pull.

The early rotative engines designed by Watt and fitted with his sun and planet gear were single acting, and in every case the beam was heavily weighted to assist the motion of the flywheel on the return stroke of the piston. In the spring of 1783 Watt experimented with a double-acting engine at Soho which incorporated a rack and sector motion designed to admit steam to either side of the piston. This was not satisfactory, and when the first full size double-acting engine was built for Coates & Jarratt of Hull in 1784, it included the parallel motion which Watt had mentioned in his patent taken out in April of that year. This was Watt's third major patent relating to the steam engine and, as in its predecessors, it contained a number of improvements of major and minor significance. These included, in addition to the parallel motion, various methods of balancing pump rods and new ways of applying steam power to rolling and slitting mills and forge hammers. The parallel or three-bar motion was the invention of which Watt himself was most proud, and without it the double-acting rotative engine would have hardly been possible. A rigid connection was required between the piston rod and its beam; this was achieved by means of a link, and Watt devised his parallel motion to guide the top of the rod in a straight line. The only major improvement made to the steam engine by Watt after this patent was the introduction of the centrifugal governor in 1788 to operate the throttle valve and control the engine speed. This type of device had already been used in windmills for regulating the distance between the grind stones (ref 33) and Watt, who knew the pitfalls of the patent law, made no attempt to claim exclusive rights for the device although it became known as the Watt governor.

Further Reading
Law, R. J., *The Steam Engine*, 1965
Dickinson, H. W., *A Short History of the Steam Engine*, 2nd edition, 1963
Rolt, L. T. C., *James Watt*, 1962

21 ROTARY STEAM ENGINE
James Watt (1736–1819)

Watt's rotary steam engine is reported to have worked satisfactorily but was too expensive and none was made for sale.

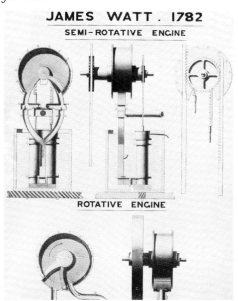

A rotary engine in which an annular chamber rotated around a central axis was devised by James Watt in 1769, and a full scale test engine was built at the Soho Works in 1774. The chamber had a diameter of 6ft and was divided into three parts by flap valves arranged at 120° to each other. The lowest part of the chamber was filled with mercury. Steam was admitted between the flap valves and the surface of the mercury causing the chamber to rotate. There were problems due to steam leakage and the high cost of mercury made the project uneconomic. Watt, and particularly Boulton, could see the advantages to be gained by achieving rotary motion, and after this initial setback Watt turned his attention to adapting the reciprocating pump engine for this purpose. The result was the five part patent of 1781 (ref 19) which included the celebrated sun and planet gear.

Further Reading
Law, R. J., *The Steam Engine*, 1965

22 STEAM CARRIAGE
William Murdock (1754–1839) and
James Watt (1736–1819)

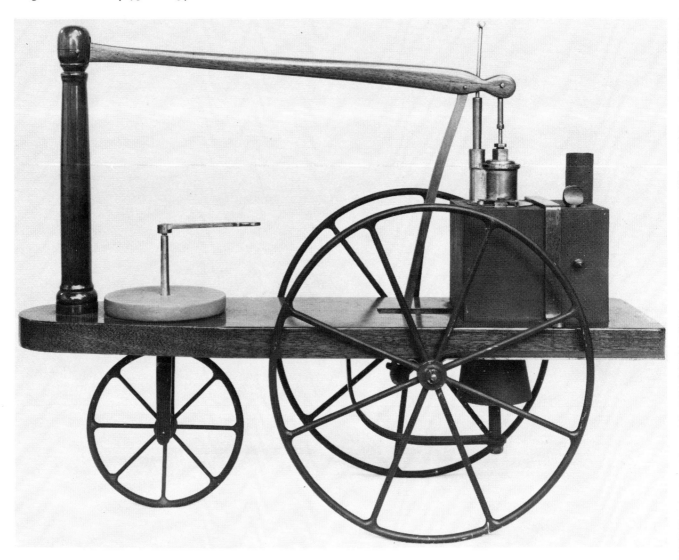

This is an example of a fascinating near miss which might have changed the course of history had the senior partner in the enterprise, James Watt, been less concerned with the production of mill and pumping engines.

William Murdock, who was for many critical years the principal erector of Boulton and Watt engines in Cornwall, was not only an engineer and inventor of the first order, but was also a man of steadfast character who displayed a loyalty towards his employers that was not always deserved. He soon became highly regarded by the Cornish mine captains and it was one of these men, John Budge, who placed a workshop and a foundry at his disposal

so that he could carry out his own experiments. Many years before, Murdock had discussed the possibility of building a steam carriage with his father but it was not until he was established in Cornwall that the opportunity materialised. He built a working model between 1781 and 1784 which was capable of drawing a model wagon at a speed of between 6 and 8 mph. He subsequently outlined his plans to Matthew Boulton of a more ambitious gear-driven locomotive with a condensing engine. Boulton undoubtedly communicated Murdock's ideas to his partner. It is not known whether Watt was already pursuing a similar line of investigation; although on the evidence of Watt's insistence on the use of low pressure rather than high pressure steam, this seems unlikely. Nevertheless Watt's patent no1432, dated 28 April 1784, included reference to the application of steam engines 'to give motion to wheel carriages'. The engine described therein showed a remarkable similarity to that in Murdock's proposals and Watt calculated that to drive a carriage containing two people, a cylinder of 7in diameter with 12in stroke would be required. Murdock must have been unaware of the patent because he continued to work on another model and in 1786 set out for London with the intention of obtaining a patent for himself. Boulton, however, intercepted him at Exeter and persuaded him to give up the project, which was probably wise since the patent application must surely have been refused. It was a few nights later when the two men were together at a hotel in Truro that the famous demonstration took place in which the model

carriage is reputed to have carried the shovel, poker and tongs from the fire grate round the room. Murdock was thereafter busily engaged on other projects (ref 126) but he is said to have made yet another model. There is an interesting link between this model and Trevithick, since the latter called at Murdock's house in Redruth to see it in 1794. Trevithick's first locomotive ran in 1804 and the Cornishman was obviously influenced by Murdock's work. Watt, regrettably, never made any attempt to exploit that part of his patent relating to the steam carriage and the world had to wait for another quarter of a century for the true steam locomotive.

Further Reading

Rolt, L. T. C., *James Watt*, 1962
Robinson, E. and Musson, A. E., *James Watt and the Steam Revolution*, 1969. *Engineering Heritage* Vol 2 1966

23 COMPOUND STEAM ENGINE
Jonathan Carter Hornblower (1753–1815)

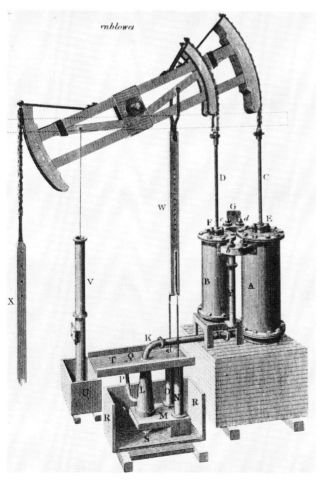

Jonathan Hornblower patented the world's first compound engine but was unable to exploit his invention fully because of the restrictions of Watt's patent for a separate condenser. The Hornblower engine, although ingenious and exhibiting a high standard of workmanship, did not benefit from the advantage of two-stage expansion because of the low initial pressure of the steam which was only just above that of the atmosphere.

Jonathan Carter Hornblower was one of the celebrated family of Cornish mining engineers whose activities spanned most of the eighteenth century, and who were collectively responsible for the erection and maintenance of many of the Newcomen and Watt pumping engines in the county. On 13 July 1781 Hornblower took out patent no1298 for a two-cylinder or compound steam engine in which the steam was admitted from the boiler to one cylinder and then passed to a second and larger cylinder where it did further work before being condensed. The first Hornblower compound engine was erected outside the inventor's home county, at Radstock Colliery near Bath, and began operating towards the end of 1782. The cylinders measured 19in and 24in respectively, and both piston rods were connected to a common overhead beam. Watt, who was also experimenting with a twin cylinder engine at the time, considered that the provision of a second cylinder in which the steam was condensed was an infringement of his separate condenser patent. For several years he confined himself to harrying his rival and his associates with threats of litigation. Hornblower continued to erect his engines in Cornwall, however, and during the years 1784 to 1791 a further eleven entered service at various mines. Eventually Boulton, Watt & Co commenced legal action but the mine owners capitulated *en masse* and agreed to pay royalties to the Midland company before the case was brought to court. Jonathan Hornblower then attempted to have his patent extended so that he might benefit from it after Watt's condenser patent expired in 1800, but this was opposed and the Cornishman retired from the fray. The compound engine was re-invented in the next century by Arthur Woolf and some other engineers, but it was not until the general adoption of high pressure steam that its full potential was successfully exploited.

Further Reading
Barton, D. B., *A History of Copper Mining in Cornwall and Devon*, Truro, 1968
Rolt, L. T. C., *James Watt*, 1962

24 STEAM ENGINE
Edward Bull (d 1799?)

The inverted cylinder type eventually became the most popular form of reciprocating steam engine, but its progress was restricted by litigation since the earliest version included a separate condenser and infringed Watt's master patent of 1775.

Edward Bull was a member of that talented band of inventor craftsmen employed by Boulton, Watt & Co as engine erectors in Cornwall and elsewhere between 1780 and 1800. Like Jonathan Hornblower and William Murdock, Bull made a notable contribution to steam engine development but was unable to exploit his invention initially to any great extent because it infringed Watt's extended patent of 1775 relating to a separate condenser. In company with other engineers who were to follow him, Bull was critical of the rather cumbersome overhead beam arrangement of the Watt engine (ref 18) which Watt had inherited from Newcomen. He avoided this by inverting the cylinder directly over the pump-rod, thus producing a direct-acting engine that was much more compact in design. Bull's first engines were built in 1792, and their merits were quickly appreciated by the Cornish mine owners who began to order the less expensive inverted cylinder type in preference to those made by Boulton & Watt. This naturally incensed the Soho partners and an action was brought against Bull, the trial taking place in London in the Court of Common Pleas in 1793. It has been stated that Watt had previously constructed inverted cylinder engines, but the main argument for the plaintiffs revolved around the alleged infringement of the separate condenser patent which was not due to expire until 1800. The names of the witnesses, both for the plaintiffs and the defendant, form almost a roll-call of eminent eighteenth-century engineers and men of science with Herschel, Ramsden, Murdock and

Cumming among those who appeared for Boulton & Watt and Bramah, Jabez Hornblower, Rowntree and Trevithick supporting Bull. The verdict was in favour of the plaintiffs, but there was an appeal in 1795 in which the judges failed to agree. Bull, however, was effectively restrained from building any more engines by a Court of Chancery injunction following the initial jury verdict. The situation altered after 1800 and a number of inverted cylinder pumping engines to Bull's design were installed in mines in the North of England where they were highly regarded.

Further Reading
Barton, D. B., *A History of Copper Mining in Cornwall and Devon*, Truro, 1968
Robinson, E. and Musson, A. E., *James Watt and the Steam Revolution*, 1969

25 STEAM PUMP
James Keir (1735-1820)

Keir's engine contributed little to general steam engine development since it was constructed on the same principle as that of Thomas Savery.

James Keir was a member of that select band of *savants*, the Lunar Society of Birmingham, which included many outstanding figures of eighteenth-century science among its membership. Educated at Edinburgh University, he first became an army physician and served in the West Indies before settling in Birmingham in 1770. He established a glass factory at Stourbridge in 1775 and for a time was manager of Boulton's Soho works at Birmingham. This appointment did not last long, for he set up in business with a former brother officer to manufacture alkali and soap at Dudley. It was probably to drive machinery in this works that his steam engine was erected in 1793. Keir's engine contained no fundamental advances; it was in fact a steam pump designed on the Savery principle (ref 12) in which water was raised from a well and led to the top of an overshot water wheel. The water caused the wheel to rotate and was returned to the well by gravity where the cycle was repeated. The practice of obtaining rotary motion by incorporating a water wheel into the engine was followed by a number of engineers, and atmospheric engines as well as Savery-type pumps were adapted in this manner. The main reason for preferring this rather devious and cumbersome arrangement was to circumvent the restrictions of Watt's various patents, particularly that of the separate condenser (ref 18) and Keir, as a friend of the Birmingham partners, was obviously unwilling to transgress in this respect. The boiler of Keir's engine was similar to Watt's rectangular tank boiler and measured 7ft in length with a width and height of 5ft. The steam vessel was located just above the height to which the water had to be raised, and, when operating at full power or ten cycles per minute, the pump would raise 70cu ft of water through a height of 20ft. The coal consumption was considerable—6 bushels (522lb) in 12 hours or 7 bushels when the boiler required cleaning. This was prodigious, even by the standards of the day, and it was estimated that Keir's engine was about 50 per cent as efficient as the Watt beam engine and even inferior in performance to the Newcomen atmospheric engine.

Further Reading
Tredgold, T., *Tredgold on the Steam Engine*, 1851

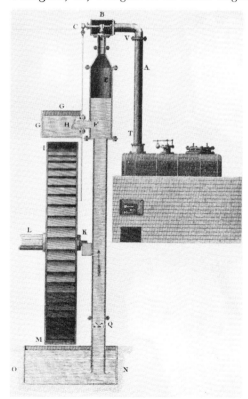

26 STEAM ENGINE
Edmund Cartwright (1743–1823)

Cartwright's steam or alcohol engine might be regarded as a scientific curiosity, but it incorporated metallic packing for the piston—the first time that this type of packing was used in steam engine construction.

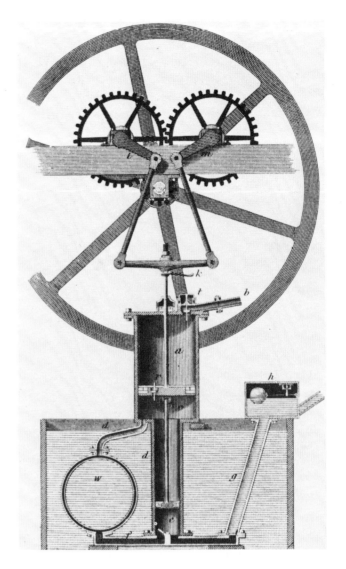

After his business in Doncaster had failed (ref 45) Edmund Cartwright moved to London, and there made the acquaintance of a number of eminent engineers including Robert Fulton. It may have been Fulton who persuaded Cartwright to interest himself in steam engines, although it was possibly inevitable that one with such a natural mechanical aptitude would concern himself with what was the most important avenue of technological development of the day. Cartwright, like other inventors of the period, could not ignore Watt's condenser patent which still had three years to run in 1797. This influenced his design. He also wished to avoid the heavy overhead beam which, although not patented, was a feature of all Newcomen and Watt engines. His solution resulted in an engine which in fact incorporated a condenser, but this was skilfully, and perhaps intentionally, hidden inside a water tank or cistern. The condenser did not include a nest of tubes but consisted of two hollow cylinders arranged concentrically. Cooling water passed through the centre of the smaller tube and around the outside of the larger tube. This presented a large cooling surface for condensing the fluid, although it is doubtful if the arrangement was more efficient than the simple jet condenser. It was in his choice of fluid that Cartwright displayed his flair for originality which had brought him so near to success with the power loom. Instead of water in the boiler, he decided to substitute alcohol which could be vaporised with much less fuel. The cycle followed the conventional path of the Watt engine, with the condensate being pumped back to the boiler by a displacement pump which was an extension to the main engine cylinder. The piston rod was also extended upwards through a stuffing box in which Cartwright fitted his metallic packing. This was not only steam-tight but, unlike the traditional hemp packing, did not require bedding in when the engine was first run up. The piston rod was connected to two cranks of equal length which were attached to two gear wheels, the latter meshing with the fly-

wheel drive. This method of converting vertical motion into reciprocating motion was not unlike Watt's sun and planet gear (ref 19). Cartwright's engine was not successful, and this may have been due to the problem of maintaining sufficient alcohol in the circuit; Cartwright is known to have considered linking the engine directly to a still, which presumably would have operated on part of the product of the process plant. It was nevertheless the first of many attempts to produce a close-cycle engine using a fluid other than steam to produce motive power.

Further Reading
Tredgold, T., *Tredgold on the Steam Engine*, 1851

27 STEAM ENGINE
Matthew Murray (1765–1826)

Matthew Murray made a number of improvements in steam engine design at the end of the eighteenth century. His engines were notable for their compactness and ease of maintenance, since the working parts were easily accessible.

As the eighteenth century drew to a close and the time approached when Watt's patent (ref 18) would expire, a number of engineers began to consider the next stages in the development of the steam engine. Among the most prominent of this group was Matthew Murray, a native of Newcastle-upon-Tyne and subsequently a partner in the firm of Fenton, Murray and Jackson, who were proprietors of the celebrated Round Foundry at Leeds. Murray was another member of that select band of engineers who rose from the ranks of the craftsmen. He was, in fact, apprenticed to a blacksmith, but moved down to Leeds in 1789, where he was employed by a firm of flax spinners. This was a period of intensive development and change in the textile industry, and he was able to exploit to the full his natural mechanical ability. He made several improvements to flax spinning machines, which ultimately led to the important advance of wet spinning, and in 1790 and 1793 he took out two patents, nos 1752 and 1971, relating to further improvements in textile machinery. It was at this time that he set up in business as a manufacturer of flax spinning machines at the Round Foundry with his two partners. The demand for steam engines to drive the mill machines was exceeding the capacity of Boulton & Watt, and a number of engineers including Murray began to look with envy at this lucrative market. Murray could also see that Watt's beam engine was considerably over designed for the duties that it had to perform. Towards the end of the century the partners, under Murray's instigation, began to build their own engines, and Murray himself took out three further patents, no 2327 in 1799, no 2531 in 1801 and no 2632 in 1802, relating to steam engine design. He is generally credited with the invention of the short D-slide valve, and is also known to have invented a planing machine to machine the face of the valve. His engines were models of compactness compared with those of his rivals in Birmingham, and he used cast iron for the overhead beams which made them stronger yet much more slender than the ponderous timber beams favoured by Watt. Murray's career as an engine builder overlapped the beginning of the railway age, and he was responsible for a number of celebrated locomotives used for the rack railways linking the Middleton Collieries with Leeds.

Further Reading
Tredgold, T., *Tredgold on the Steam Engine*, 1851

WATER AND WIND POWER

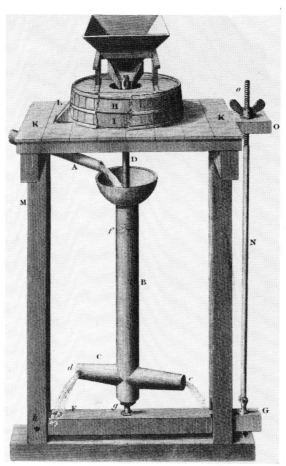

28 WATER TURBINE
Robert Barker

Although never put to any great practical use, Barker's mill incorporated an early form of water turbine which operated on the reaction principle similar to a modern garden sprinkler.

Robert Barker's mill, which was first constructed in 1745, was an ingenious attempt to produce an apparatus for grinding corn that would be smaller and cheaper to make than the standard breast water mill. All that was required was a continuous head of water to supply the motive force. Water was fed by gravity from a pipe into the central vertical column of the mill, and discharged from two holes drilled at the ends of a pair of nozzles set at 180° to each other. The reaction created by the jets emerging from the holes was sufficient to turn the central column which rotated the upper stone of a pair of grindstones, located above the vertical column. This part of the apparatus was almost identical to that fitted in the common breast mill, and a hopper was provided through which to feed the corn. There is little evidence to show that the mill was put to much practical use but it is known to have intrigued many leading scientists and engineers. Desaguliers, Euler and John Bernouilli all examined the machine and improvements were suggested by Mathou de la Cour in 1775. Despite improvements in windmills and the introduction of other forms of rotary drive, the mill was not completely ignored in the nineteenth century and in 1845 James Whitelaw of Paisley took out a patent for further improvements relating mainly to the arrangement of the nozzles. These were shaped in the form of a curve and water was discharged from the ends, instead of through a hole, at 90° to the axis.

Further Reading
Glynn, J., *A Treatise on Water Power*, 1853

29 AUTOMATIC FANTAIL FOR WINDMILLS
Edmund Lee

The windmill fantail invented by Edmund Lee was an early example of automatic control with feed back. Certain earlier windmills were winched round into the wind but Lee's device was self-regulating and considerably improved operational efficiency.

The early post mills of the sixteenth and seventeenth centuries were adapted so that they could be rotated by means of a winch to ensure that the sails were always turned into the wind. This was a time-consuming process and difficult to apply to the larger tower mills which superseded the post mills The problem was solved by Edmund Lee in 1745 with his invention of the automatic fantail. This consisted of a number of vanes set at right angles to the sails of the mill. The position of the fantail varied according to the type of mill but in the tower mill illustrated it is mounted at the back of the cap. The operation was essentially simple. The fantail remained stationary as long as the sails were facing into the wind, but when the direction of the wind changed it caused the fantail to rotate which in turn moved the cap round through a system of bevel gearing. Since the main sails were also mounted on the cap, these also moved round until they were once more back into the eye of the wind. Post mills were also fitted with fantails but a different arrangement was adopted with both the mill and the ladder running round separate tracks on the ground.

30 IMPROVEMENTS TO WATER WHEELS
John Smeaton (1724–92)

As a result of Smeaton's experimental work water mills became more efficient and their lease of life was considerably extended, particularly in areas away from coalfields where they could even compete effectively with the Watt beam engine.

The water mill was an established source of power long before the eighteenth century, but it reached its peak of development during the latter years of the century and in the immediate pre-Victorian era before the steam revolution was consolidated. John Smeaton was probably the first man to discard the empirical approach to water wheel design. He constructed a model test rig which included a water tower, a hand pump and a mechanism for raising a weight to demonstrate the wheel's efficiency. Provision was made for adjusting the angles of the wheel blades, and the apparatus could be used for both overshot and undershot wheels. Smeaton's experiments proved conclusively that the highest efficiency would result when the buckets were filled with water and the wheel turned by gravity; these conditions were obtained in overshot wheels, which proved in the model tests to be more than twice as efficient as the undershot type. Smeaton was no sterile theoretician; during his life he built forty-three water mills and introduced a number of practical innovations. His most celebrated mill was perhaps the one installed at the ironworks of the Carron Company in 1769. This wheel, which operated a furnace blowing engine, was the first to be equipped with a cast iron shaft. Later Smeaton mills were also fitted with cast iron gear wheels and, although certain production problems were encountered mainly due to porosity in parts of the castings, this innovation was generally adopted.

Further Reading
Engineering Heritage Vol 2, 1966

Further Reading
Wailes, R, 'Windmills' *Engineering Heritage* Vol 2, 1966

31 TIDAL PUMP
John Smeaton (1724–92)

John Smeaton's tidal pump at London Bridge was the last of several such machines which supplied water to subscribers in the City of London in the seventeenth and eighteenth centuries.

The first tidal pump to be built at London Bridge during the eighteenth century was that of George Sorocold in 1702. This was constructed for the newly-formed London Bridge Water Company which undertook to supply water to subscribers in the City of London at a fixed charge. In 1767 the second arch at the south end of the bridge was provided with another pumping plant by John Smeaton. His design was similar to that of Sorocold; it was however more compact, with one large wheel known as the Borough Wheel substituted for the three arranged in parallel by Sorocold. Like the earlier machine Smeaton's pump was constructed

principally of timber. The maximum diameter of
the wheel was 32ft, and it carried twenty-four vanes
or floats which took the impulse of the water flow
between the arch of the bridge. Two spur wheels of
15ft diameter were arranged on either side of the
main shaft. These engaged pinions, which in turn
operated three cranks set at 120°. The connecting
rods attached to the cranks were made of wrought
iron. At their upper ends they were linked to
massive overhead beams which pivoted about their
mid-points. The other ends of the beams operated
the pump rods. Six pumps were fitted, each having
a diameter of 10in and a 4½ft stroke. The tidal
pump worked satisfactorily, except in dry summers
or when the tide was unusually low. An atmospheric
engine was installed to work in conjunction with the
water wheel, and Smeaton reported on its operation
to the water company in 1771. The water pumps

discharged to a wooden water tower, but this was
destroyed by fire in 1779. The engine was un-
damaged, and the service was resumed in a few
days with the discharge being led straight to the
mains. The wooden wheel was replaced by one of
iron in the nineteenth century, and the pump was
finally dismantled when the old bridge gave way
to Rennie's new stone bridge in 1831. By that time
the New River Company had supplanted the London
Bridge Company and taken over its obligation to
consumers in the City.

Further Reading
Panell, J. P. M., *An Illustrated History of Civil
Engineering*, 1964
Armytage, W. H. G., *A Social History of Engineering*,
1961

32 WATER PRESSURE ENGINE
John Smeaton (1724–92)

*The water pressure engine was developed for pumping out
mine workings as an alternative to the atmospheric steam
engine; it could not compete with the more efficient Watt
condensing engine and few examples were made after the end
of the eighteenth century.*

The coal consumption of the atmospheric steam
engine was so large that this type of engine was only
economically suitable for use in coal mines where
there were ample stocks of coal on hand. Owners of
copper and lead mines, where the pits were sunk

progressively deeper into the earth as the demand for these metals increased, were faced with ever rising costs in their efforts to prevent the mine workings from flooding. It was against this background that the first water pressure engine was built in England by William Westgarth to pump out water from a lead mine in Northumberland in 1765. Westgarth was not the true inventor of this type of engine as it was also used on the continent in Germany and Hungary before this date. Westgarth, on the advice of John Smeaton, applied to the Society of Arts for a grant to develop his invention but did not take out a patent. He was asked to make a model and submit plans to the society but died before these requirements could be satisfied. Smeaton, who had previously suggested various improvements to Westgarth, then took over the project and constructed a water pressure engine at Temple Newsam in Yorkshire to pump water to a house owned by Lord Irwin. Although the principle of the water pressure engine was simple, the mechanism was fairly complex. A constant head of water was required and water was led by gravity to the cylinder to which was fitted a slide valve of ingenious design. This comprised a circular hoop or ring which was permitted to slide for a short distance up and down the inlet pipe. When the upper parts were exposed, water entered the cylinder below the piston and caused it to rise. The piston rod was linked to an overhead beam which was pivoted at its fixed end. The beam was itself linked to the pump rod and movement of the beam produced a corresponding movement of the pump

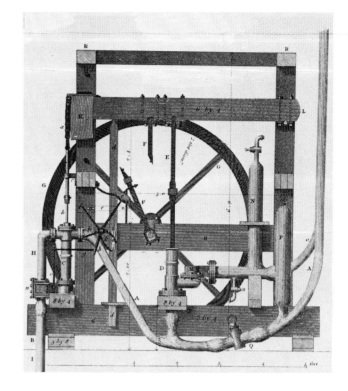

plunger. Smeaton's engine preceded the Watt condensing engine by a mere half dozen years and it could not compete with this more efficient version of the steam engine. It was revived briefly by Trevithick after Smeaton's death but was again superseded by more flexible forms of water power developed during the nineteenth century.

Further Reading
Glynn, J., *A Treatise on Water Power*, 1853

33 CENTRIFUGAL GOVERNOR FOR WINDMILLS
Thomas Mead

Following the introduction of the centrifugal governor, millers were able to maintain grinding conditions fairly constant and hence produce particles of uniform size.

Although centrifugal speed governors of the rotating ball type were probably familiar to eighteenth-century millwrights, it was not until 1783 that Thomas Mead patented the application of this device for controlling the speeds of millstones. In Mead's patent no1628 the governor was mounted on the same shaft as the top or rotating grindstone. As the speed increased the top grindstone, under

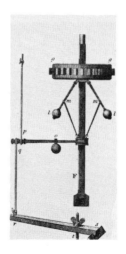

normal conditions, would tend to rise and cause unevenness and irregularity of the flour particle size. Prior to the introduction of the governor the miller, in gusty conditions, had to control the gap between the two stones manually by means of a cord and a system of levers. However, with the governor an outward movement of the rotating weights as the speed increased actuated the levers automatically and the gap between the two stones was held at a predetermined level.

Further Reading
Wailes, R. 'Windmills', *Engineering Heritage* Vol I, 1963

34 AUTOMATIC FLOUR MILL
Oliver Evans (1756–1819)

Oliver Evans was the pioneer of continuous flow automation and, although his ideas received scant attention during his own lifetime, they exerted a far reaching influence in the USA in the nineteenth century, particularly as the tempo of industrialisation accelerated.

Oliver Evans was the fifth of a Delaware farmer's twelve children. In 1780 towards the end of the Revolutionary War, Evans and several of his brothers set up in business as flour millers at Wilmington, Delaware. Labour was so scarce that Evans decided that the only solution was to eliminate all manual stages in the milling operation. The source of power was a conventional water wheel and this was linked through a system of simple gears not only to the grind stones, but to a number of devices which reduced handling within the mill to a minimum. The corn was first received at the mill from a wagon or boat and raised to the upper storey by an elevator where it was cleaned and, if necessary, put into store. When required for milling, it was fed through a series of hoppers to the millstones. After grinding it was returned to the top on another elevator, where it was successively cooled and sieved to separate the bran and partly ground meal from the flour. The final operation was performed by several horizontal screw conveyors which moved the flour to the bagging store. Evans' ideas were far ahead of his time. In 1790 he received a Congressional Patent—one of the earliest American patents—and five years later he published details of his inventions in a book entitled *The Young Millwright and Miller's Guide*. The flour millers, however, remained sceptical and although Joseph Evans, one of the brothers, travelled through Delaware, Pennsylvania, Maryland and Virginia attempting to sell the idea of an automated mill, no interest was shown. Evans then induced a friend who was visiting England to approach millers on the other side of the Atlantic but again there was no response. The English mills at the time were smaller and far more numerous than their American counterparts, and few millers, even if they saw merit in Evans' proposals, could afford the capital outlay required. Eventually the climate of opinion changed in the USA, but Evans made little profit out of his invention. Numerous violations of his patent were discovered, which resulted in a number of costly lawsuits, as well as many others that went undetected. Evans was nevertheless one of the true founders of American engineering and his versatile genius manifested itself in many other directions during the early years of the nineteenth century—notably in steam propulsion where he was a contemporary and rival of Trevithick.

Further Reading
Cummings, A. D., 'Oliver Evans—Pioneer of Automation', *Engineering Heritage* Vol 2, 1966

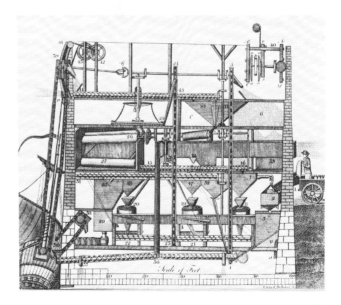

35 HORIZONTAL WINDMILL
Stephen Hooper

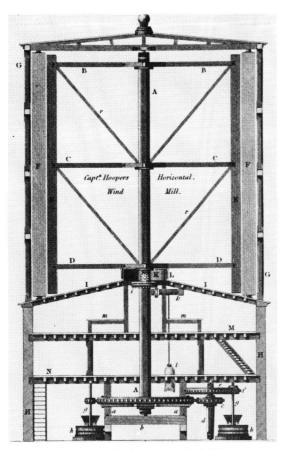

Stephen Hooper's horizontal windmills were no more successful than others of similar design built in England during the eighteenth and nineteenth centuries, but his roller reefing sails, which prevented excessive speeds, were an important development in the latter-day evolution of the windmill.

Windmills, in which the sails were mounted on a vertical windshaft and revolved in a horizontal plane, existed in Persia as early as the tenth century AD. This type of mill long intrigued European windmill builders, and towards the end of the eighteenth century three notable examples were constructed by Stephen Hooper of Margate. Hooper's first horizontal mill was built on his property at Margate, and consisted of a mill tower above which was erected a large wheel, of about 28ft diameter and 48ft high, that revolved in a horizontal plane. The wheel was fitted with vanes like a paddle-wheel and was surrounded by a concentric casing formed of slats which could be adjusted to act on one part of the wheel while the remainder of it was screened.

The shaft on which the wheel was keyed drove five pairs of millstones in the conventional way. Hooper's second horizontal mill was built in London on the present site of Battersea Park. An elevation and plan of this mill are shown in the illustration; the base was octagonal and the total height including the wheel was 145ft. The outer casing included ninety-six movable shutters, and the wheel contained the same number of vanes. The mill was originally operated by a maltster and was used for grinding linseed; it required regular maintenance and eventually the cost of its upkeep was such that it fell into disrepair and was finally dismantled in 1849 when Battersea Park was laid out. Towards the end of the century, Hooper was employed on the fortifications at Sheerness and built his third horizontal mill to draw water from a well. As an alternative source of power, arrangements were made to use horses in calm weather. A more lasting contribution by Hooper to windmill development was his roller reefing sail which he patented in 1789. He used a number of small roller blinds in place of the usual shutters which could be opened and closed simultaneously. It was his intention that the action should be automatic and prevent excessive speeds in high winds, but in practice some degree of manual operation was necessary.

Further Reading
Wailes, R., *The English Windmill*, 1954

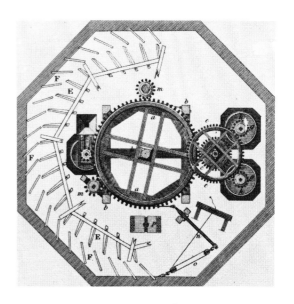

TEXTILES

36 SILK MANUFACTURE
Thomas Lombe (1685–1739)

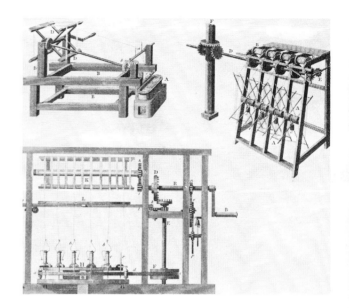

The mechanisation of silk spinning led to the erection of a mill which was one of the earliest examples of the factory system in Britain. Lombe's silk spinning machinery was a major achievement although it owed much to similar Italian machines, and there is evidence that he and his brother resorted to industrial espionage to obtain the secrets of their continental rivals.

Until the beginning of the eighteenth century, Italian manufacturers dominated the silk industry in Europe and jealously guarded the secrets of their trade. Their monopoly was broken by an enterprising London merchant, Thomas Lombe, who was himself the son of a worsted weaver of Norwich. Lombe sent his half-brother John to Italy to study the Italian methods of spinning. The information that he brought back, coupled with Thomas Lombe's manual dexterity and inherited knowledge of the principles of weaving, enabled the brothers to reproduce and improve upon the Italian machinery. In 1718 Thomas Lombe was granted a patent, no422, for the invention of 'three sorts of engines never before made or used in Great Britain, one to wind the finest raw silk, another to spin and another to twist the finest Italian raw silk into organzine in great perfection, which has never before been done in this country'. A year after the patent was granted the brothers set up a mill at Derby on an island in the River Derwent. This was one of the first examples of the factory system as applied to the textile industry and anticipated the methods of Arkwright and Strutt by fifty years. The enterprise eventually became a success, and the mill remained in operation on that site for a hundred and fifty years.

Lombe, as the founder of the English silk spinning industry, received a reasonable remuneration for his work, although his patent was not extended beyond the original fourteen years despite an application supported by fellow merchants in the City. Instead he was awarded the sum of £14,000 on provision that he should display models of his machinery in a suitable public place—the Tower of London. Lombe, like Arkwright, was a man who combined and inventive mind with considerable business acumen. His affairs continued to prosper as he grew older; he was knighted in 1727 and when he died left £120,000—a substantial sum for the day.

Further Reading
Lewis, J., *The Silk Book*, 1951

37 COTTON SPINNING MACHINE
Lewis Paul (d 1759) and
John Wyatt (1700–66)

The spinning machine of Paul and Wyatt preceded that of Arkwright by over thirty years but a variety of factors, including high freight costs to and from the Birmingham area, prevented it from becoming a commercial success.

Considerable controversy has always centred around the invention of the first cotton spinning machine which was the subject of patent no562

awarded on 20 July 1738 to Lewis Paul, then described as a gentleman of Birmingham in the County of Warwick. Paul had been left as a boy under the guardianship of Lord Shaftesbury; he showed a natural mechanical talent and in 1730 invented a machine for pinking shrouds, from which he made a considerable profit. He subsequently became associated with John Wyatt, a carpenter

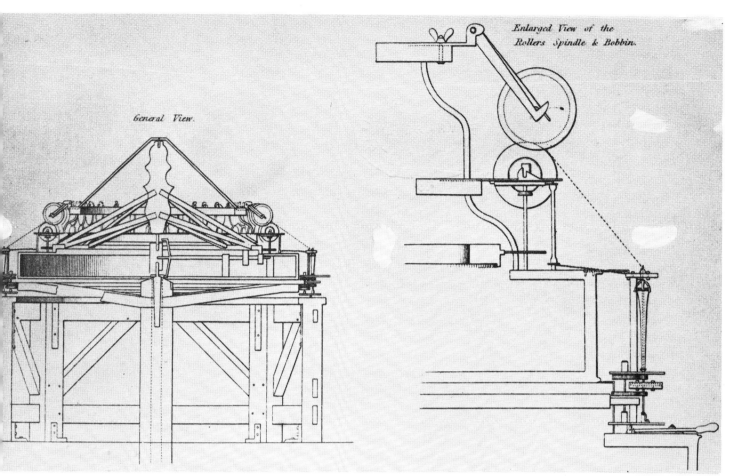

General View.

Enlarged View of the
Rollers Spindle & Bobbin.

who had attempted unsuccessfully to make a machine for manufacturing files. There are many textile historians who regard Wyatt as the true inventor of the spinning machine although it was Paul's name that appeared on the patent, Wyatt only signing as a witness. The machine introduced the principle of roller spinning which Arkwright was to develop so successfully a generation later. It consisted essentially of two pairs of rollers, one pair revolving at a slightly greater speed than the other. Slivers of cotton or wool were passed through the rollers and stretched by the action of the second pair of rollers which tended to pull the sliver faster than the first pair could deliver it. A small degree of twist was also imparted at the same time. A mill was set up in Birmingham and the enterprise received financial support from Thomas Warren, a well-known Birmingham printer, and Edward Cave who was editor of the *Gentleman's Magazine*. Paul appears to have assumed the role of superintendent at the mill and Wyatt spent most of his time in London attempting to sell the yarn. A second mill was started at Northampton, which was powered by a water mill, unlike the machinery at Birmingham where the motive power had been supplied by two donkeys. The Northampton factory contained 250 spindles and employed fifty people but, like its predecessor in Birmingham, did not prosper. One reason advanced for the failure of the enterprise was the difficulty in transporting the raw cotton and the finished yarn to and from the works. Wyatt and Paul subsequently separated; Wyatt became an employee of Matthew Boulton and invented a successful weighing machine while Paul continued to concern himself with textile machinery. He patented a carding machine (ref 40) in 1748 and in 1758, a year before he died, he took out another patent, no724, for a spinning machine, illustrated above. He endeavoured to get this machine introduced into the Foundling Hospital and a letter was drafted on his behalf by Dr Johnson who had been associated with the Birmingham enterprise.

Further Reading
Baines, E., *The History of the Cotton Manufacture in England*, 1835

47

38 FLYING SHUTTLE
John Kay (1704–64)

The flying shuttle permitted a weaver to weave more quickly with less effort and to dispense with assistants on wider looms.

John Kay was a native of Bury in Lancashire, but it was while he was employed as a hand loom weaver at Colchester in 1738 that he devised the flying shuttle. In the traditional method of weaving the shuttle was thrown from side to side by hand, a laborious process which required an additional weaver or apprentice when wide cloth was produced. Kay extended the lathe in which the shuttle ran by about a foot on each side and attached a pair of strings to the actuating mechanism. The other ends of the strings were joined to a picking stick held in the weaver's hand, and by tugging the appropriate string the shuttle was despatched across the loom into the opposite box. Kay returned to Bury where his invention was adopted by the woollen industry in the town. The cotton weavers were more reluctant to change, and it was not until about 1760 that the flying shuttle was used to any extent for the production of cotton or linen cloth. In that year Kay's son Robert invented the drop-box, which enabled the weaver to use any one of three shuttles, each containing a different colour yarn, without continually removing and replacing them in the lathe. John Kay's flying shuttle is said to have nearly doubled the production of cloth. It created an increased demand for yarn, but its adoption was bitterly opposed by many weavers who feared the loss of their employment. Kay himself was persecuted and eventually went to live in Paris.

Further Reading
Baines, E., *The History of the Cotton Manufacture in England*, 1835
English, W., 'Spinning and Weaving', *Engineering Heritage*, Vol 2, 1966

39 DRAW LOOM
Jacques Vaucanson (1709–82)

Vaucanson's loom was one of several invented in France during the eighteenth century for weaving intricate patterns in silk.

Although Jacques Vaucanson first gained fame as an inventor of automata, such as his celebrated mechanical duck, his natural mechanical aptitude made him one of the most versatile of French inventors during the middle decades of the eighteenth century. His draw loom used for silk weaving is thought to date from about 1747. It was not an original invention but rather a development of earlier looms which in turn were based on the draw boy, a fifteenth-century French invention. In this device groups of cords required to produce a pattern in the fabric were connected by a single master cord. Other advances included the use of a punched cylinder of paper to select threads of the warp to be raised and lowered. Vaucanson incorporated these and other improvements into a single machine. Several models were made but their complexity and consequent high cost prevented their widespread adoption; one machine which was bought for the city of Tours cost 100,000 livres. Like the textile machinery inventors in Britain, Vaucanson was abused and persecuted by weavers who feared that their livelihood was threatened. The fears were probably groundless and although some Vaucanson looms were actually imported into England in 1765, when a Mr Garside of Manchester installed several in a mill, each loom still required an individual operator. Vaucanson's original model loom was deposited in the Conservatoire des Arts et Métiers in Paris where it still remains. At the turn of the century it was studied by Jacquard and many of its more important features were incorporated in the famous Jacquard loom which was to become one of the cornerstones of the modern silk weaving industry.

Further Reading
McCloy, S. T., *French Inventions of the Eighteenth Century*, Kentucky, 1952
Kranzberg, M. and Pursell, Carroll W., Jr, *Technology in Western Civilisation*, Vol 1, New York, 1967

40 CARDING MACHINE
Lewis Paul (d 1759)

Paul's carding machine was the direct forerunner of those introduced into the Lancashire cotton industry in the last quarter of the eighteenth century which were based on the patents of Arkwright and others. It made possible the eventual mechanisation of another stage in cloth manufacture.

The spinning machines of Paul and Wyatt could consume the slivers of cotton for spinning faster than the carders using the hand carding technique could supply them. To overcome this limitation in output,

Lewis Paul patented the first carding machine in August 1748. The patent, no636, included a description of two different machines for accomplishing the same purpose, namely to disentangle and straighten the fibres to form slivers ready for spinning. Both machines worked on the same principle, but the cylinder machine only is illustrated. The cylinder, which was manually operated, was arranged horizontally with the entire surface area covered with cards. These were virtually identical with the traditional hand cards, and each comprised a wire

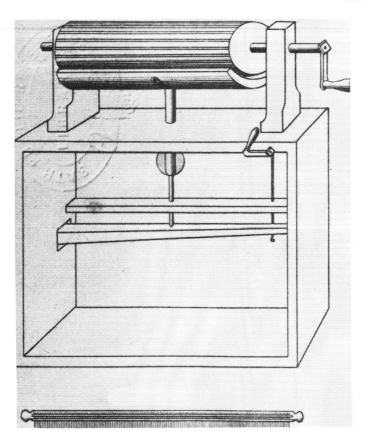

brush with the wires inclined at a pre-determined angle. Underneath the cylinder was a concave frame similarly lined with cards so that when the cylinder revolved the action of the two sets of cards combed and straightened the cotton or wool fibres in the same manner as the hand carders accomplished with two separate cards. A needle stick was provided for combing or stripping off the slivers from the cards. Paul's machine appears to have been in service for some years until the establishment at Northampton was closed. It was then bought by a Mr Morris and taken to Leominster, where it was used for carding wool for hat-making. In 1760 it was introduced to Lancashire and reverted to carding of cotton in a mill owned by Morris near Wigan. When the patent rights expired Paul's carding machine was copied and improved by men such as Hargreaves, Lees and Arkwright, and soon became a standard item of equipment in the booming textile industry of Lancashire and Scotland.

Further Reading
Baines, E., *A History of the Cotton Manufacture in England*, 1835

41 SPINNING JENNY
James Hargreaves (d 1778)

The Jenny was one of several machines which helped to transform textile spinning and weaving from a cottage industry to one that was eventually factory-based. It was used on a considerable scale in the woollen industry where it increased the supply of yarn available for weaving by upwards of a hundredfold.

James Hargreaves was one of the small select band of craftsmen who revolutionised the textile industry in Britain during the second half of the eighteeth century. A native of Blackburn, Hargreaves began his working life as a carpenter and handloom weaver. In 1760 he made a carding machine and a few years later in or about 1764 he invented the spinning jenny which was in effect a multiple spinning wheel. Hargreaves, in fact, was supposed to have conceived the idea for such a machine through watching an ordinary spinning wheel continue to revolve after it had fallen on its side. It occurred to him that if a number of spindles were placed upright side by side, it would be possible to spin several threads at once. In his first

machine he incorporated eight spindles thus increasing the output of the operator eightfold but later machines were built with up to a hundred and twenty spindles. The operator was able to cope with the vastly increased number of spindles because the fibres were fed in through a single clamp which could be operated by one hand while the other hand was used to turn the handwheel. The jenny was widely used in the woollen industry where it was employed for warp and weft production. Hargreaves, however, did not benefit greatly from his invention. His first machine was used by his family but when he began to make other machines for sale in 1768 the other spinners in Blackburn burnt his house down and destroyed the jenny. As a result Hargreaves moved to Nottingham and entered into partnership with a Mr James who built a small mill in which the jenny was used. It was while he was living in Nottingham that Hargreaves patented his machine but this did not prevent some Lancashire cotton manufacturers from constructing similar machines of their own. The last years of

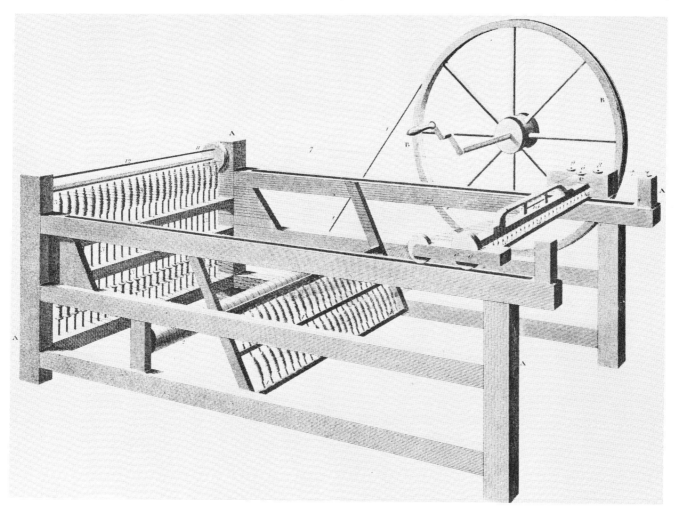

Hargreaves' life until he died in 1778 were occupied to a large extent with lawsuits relating to his patent, few of which were resolved to his satisfaction.

42 WATER FRAME
Richard Arkwright (1732–92)

Apart from Watt's improvements to the steam engine, Arkwright's spinning frame was probably the most significant invention of the eighteenth century. It not only accelerated the introduction of the factory system in the textile industry, but it complemented the inventions of Kay and others, and ensured an almost inexhaustible supply of high quality yarn for weaving.

Richard Arkwright was in succession a barber, publican and travelling wig-maker before he became involved in the problems of the mechanical spinning of cotton. In 1761 Arkwright married Margret Biggins of Leigh in Lancashire. Leigh was also the home town of Thomas Highs who, with the assistance

Further Reading

English, W., 'Spinning and Weaving', *Engineering Heritage*, Vol 1, 1963

of a clockmaker named Kay, had built an experimental spinning machine. Arkwright and Kay subsequently joined forces and together made a number of models; these must have been promising, for they obtained financial backing and moved to Preston, where they rented a room in a house attached to the Preston Free Grammar School to continue their experiments. Despite attempts at secrecy, their intentions became known to those engaged in the local cottage industry of hand spinning, and Arkwright wisely decided to move to Nottingham before he received the same treatment as Hargreaves, the inventor of the spinning jenny (ref 41). It was at Nottingham that he met Jedebiah

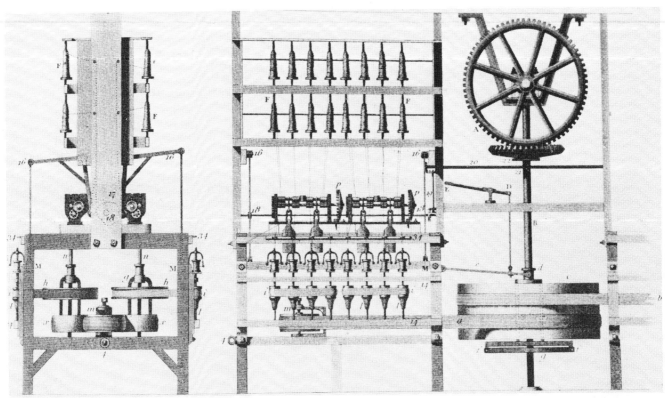

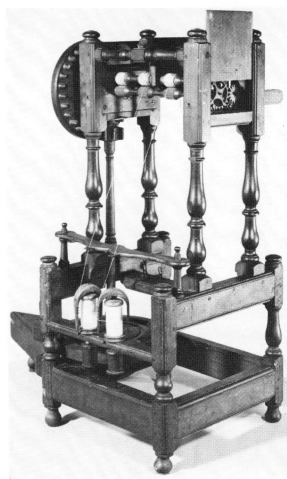

Strutt, who had previously invented a machine for manufacturing ribbed stockings. Strutt immediately realised the possibilities of Arkwright's machine, for which the latter had obtained patent no931 on 3 July 1769. A few machines had previously been erected in a small mill in Nottingham which were worked by horse power, but Strutt, who became a legal partner of Arkwright, erected a larger mill in 1771 on the banks of the river Derwent in Derbyshire. A second mill was built soon afterwards at Belper, and Arkwright continued working on the machine to improve its operation and to mechanise the processes which preceded the actual spinning operation. He was granted a second patent, no1111, on 10 December 1775 for a carding machine for preparing cotton, silk, flax and wool for spinning. The water frame, as the spinning machine became known when driven by a water mill, was however the key invention. His original model of 1769 consisted of four pairs of rollers. The top roller in each pair was covered with leather to grip the cotton, and the lower roller was fluted internally to allow the cotton to pass through. One pair of rollers revolved faster than the rest, causing the cotton sliver or rove to be drawn out to the required fineness. Twisting was carried out by spindles placed in front of each set of rollers. This method of roller spinning was similar to that patented by Lewis Paul (ref 37) in 1738, and it is possible that Arkwright may have heard of Paul's machine. The quality of the yarn produced by the water frame was consistently high, and gradually Arkwright's business began to prosper. In 1781 his partnership with Strutt was dissolved; Arkwright however continued to build other mills and sold the rights of his patents to other spinners. He was knighted in 1786, and died a very wealthy man at Willersley Castle which he had built at Cromford. The photograph shows Arkwright's original machine as patented in 1769, while the drawings date from 1813 and show the rapid development that had taken place in the intervening period.

Further Reading

English, W., 'Richard Arkwright—Inventor and Organiser' *Engineering Heritage*, Vol 2, 1966

43 SPINNING MULE
Samuel Crompton (1753-1827)

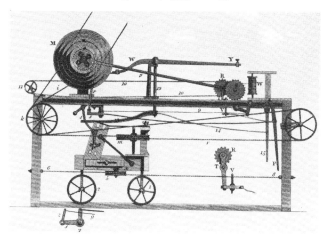

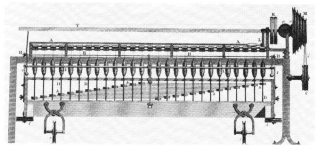

The spinning mule, like the jenny and the water frame, increased the supply of yarn available to weavers, and the quality of the yarn was superior to that produced by other types of spinning machine.

Samuel Crompton's invention of the spinning mule came at a singularly appropriate time in the development of the textile industry. By 1770 Kay's flying shuttle had become a standard attachment to handlooms, and there was a dire need for increased yarn production. Hargreaves' spinning jenny partly fulfilled this need but many operators found it difficult to control and the quality of the yarn suffered. Crompton was familiar with the jenny and he incorporated certain of its features, notably the multiple plain spindles, in his first machine which he built over a period of eight years between 1772–80. He also used drawing rollers similar to those incorporated in Arkwright's water frame, and it was possibly this fact which deterred him from taking out a patent for the mule. Crompton's first machine had 48 spindles and these were mounted on a movable carriage. The latter travelled away from the rollers, usually at a slightly faster rate than the delivery rate of the last pair of rollers. The method of winding the yarn on to the spindles was similar to that adopted by Hargreaves, but it was the inclusion of the movable carriage which enabled the mule to produce yarn that was much finer than the product of the jenny or the water frame. The machine in fact became known as the 'muslin wheel' because weavers were able to use the fine yarn it produced for muslins instead of importing hand-spun yarns from India. The mule brought fortune to many in the cotton industry but Crompton, without the protection of a patent, suffered from the avarice of those who unfairly exploited his invention. He agreed to allow manufacturers in his native Bolton and the surrounding district to make copies of his machine in return for fair remuneration. This only amounted to £60, although about ten years later a group of Manchester businessmen raised a further sum of between £400 and £500. In 1812 a petition was presented to parliament on his behalf; James Watt, who was among those who supported the petition, told the committee that in the Lancashire cotton industry mules powered by steam engines from Boulton, Watt & Co outnumbered all other types of machine. Crompton was eventually awarded £5,000, but with his lack of business acumen he derived little benefit from the money and would have experienced real poverty in his last years had a few of his friends not purchased an annuity for him.

Further Reading
English, W., 'Samuel Crompton—a failure in business', *Engineering Heritage*, Vol 1, 1963
Cameron, H. C., *Samuel Crompton*, 1951

44 TEXTILE PRINTING
Thomas Bell

The introduction of the cylinder printing of textiles, particularly calicoes, was an advance comparable in importance with that of the spinning frame, and the ready availability of high quality printing cylinders in Britain gave the British cotton industry a major advantage over its foreign competitors.

The original centre for the printing of textiles, such as cotton and linen goods, was in the London area. The process was carried out by hand using blocks of sycamore on which the pattern was engraved on one side. The block, which measured about 10in by 5in, was first brought into contact with the dye and then applied to the cloth, wire points being provided at the corners for accurate location. A tap on the back with an iron mallet imparted the pattern and the process was repeated until the entire surface was decorated. It was a slow, tedious operation, and incapable of producing intricate patterns or designs. Some improvement was effected with the introduction of flat copper plates but the output was still low. Several inventors attempted to introduce roller printing in the eighteenth century and as early as 1743 a patent was taken out by Keen and Platt for a three-colour roller process but it was not developed commercially. In France J. A. Bonvallet experimented with a similar machine at Amiens in 1775 but the first commercially successful patents were those granted to a Scotsman, Thomas Bell, in 1783. Bell's patents, no1378 and no1483, described a method of colour printing involving one to five colours applied in succession on the same machine. The key component was an engraved copper cylinder, 3–4in in diameter and slightly wider than the cloth. The lower part of the cylinder revolved in the colour bath and the excess colour was removed by a steel doctor blade before the printing operation took place. Provision was made for additional cylinders to print other colours in succession before the cloth was dried by passing over steam-heated boxes. At a single stroke, Bell mechanised textile printing and the increase in

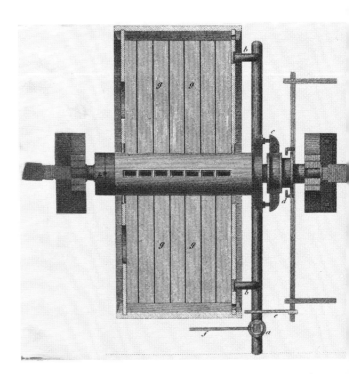

output was tremendous with a single machine, attended by one man and a boy, being capable of producing the same amount of finished work as one hundred block printers. The first installation of a cylinder printing machine was at Mosney near Preston in 1784–5 at the mill of Livesey, Hargreaves, Hall and Co. The technique spread rapidly throughout the industry and further improvements were made in the early years of the nineteenth century by men such as Joseph Lockett of Manchester and Jacob Perkins, an American domiciled in London. Block printing continued to be practised for some years, but by the 1830s three quarters of all printed textiles produced in Britain were machine printed, and British-made printing rolls were exported all over the world.

Further Reading
Baines, E., *The History of the Cotton Manufacture in England*, 1835

45 POWER LOOM
Edmund Cartwright (1743-1823)

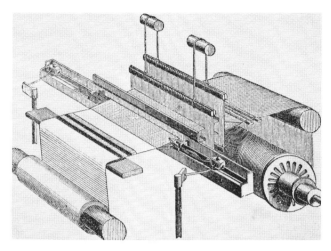

Although Cartwright's power loom did not bring him any direct financial benefits, it was the true forerunner of the machines which were rapidly adopted by the textile industry in the early years of the nineteenth century.

Edmund Cartwright could perhaps be described as a latter-day Renaissance man. A fellow of Magdalen College, a Doctor of Divinity and an accomplished poet, he was above all a talented inventor with a flair and versatility that were matched among his contemporaries only by Bramah and Smeaton. He was forty-one years old before he applied himself seriously to mechanical problems, and yet in a few short years he invented a complex machine—the power loom—which made possible the tremendous expansion in the textile industry that occurred in the nineteenth century when the Industrial Revolution gained momentum. Cartwright's power loom was designed to meet the immediate needs of weavers who were faced with seemingly inexhaustible supplies of thread and yarn from Arkwright's water frame and the numerous versions of the spinning machine which were expected to appear when Arkwright's patent expired. Cartwright's first power loom patent, no1470, was

taken out in April 1785, but the machine was clumsy and inadequate. Further patents incorporating considerable improvements were taken out in 1786 and 1787, and the resulting machine exhibited many features that were to form the basis of the early commercial power looms. The main problem was to convert the rotary motion of the water wheel into three different forms of reciprocating motion, one of which was required to simulate the motion of the hand-loom weaver in propelling the shuttle to and fro across the loom. This proved the most difficult to reproduce, but Cartwright eventually used a spring which was designed to actuate a picker stick; the latter was returned to the striking position by a ram, thus re-loading the spring. Cartwright set up a factory at Doncaster, where he installed a number of power looms and also introduced a steam engine to provide the necessary power. The enterprise was a failure, but this was probably due to Cartwright's lack of business experience rather than any fundamental mechanical deficiency of the looms. In 1790 a Lancashire manufacturer named Grimshaw installed a number of Cartwright's power looms in a factory at Gorton, and these are said to have performed satisfactorily until the factory was burnt down a few years later. Cartwright moved to London in 1793 and abandoned further work on the power loom, having already spent a good deal of money on the project. His patent expired in 1804, and two years later he petitioned parliament as the machine with further improvements was coming into general use. Fifty textile companies supported his claim and eventually in 1809 the House of Commons voted him £10,000.

Further Reading

Baines, E., *The History of Cotton Manufacture in England*, 1835

English, W., *Engineering Heritage*, Vol 2, 1966

46 COTTON GIN
Eli Whitney (1765-1825)

The invention of the cotton gin came at a most opportune moment and ensured adequate supplies of high quality, short stapled American cotton for the Lancashire industry.

Eli Whitney was born on a farm at Westborough, Massachussetts, and attended Yale College from which he graduated in 1792. He realised the possibilities of expansion of the American cotton

industry, and moved to Savannah in Georgia where he came into contact with a group of cotton growers and merchants who were seeking a means of separating the short staple upland cotton from its seeds. After only a few weeks Whitney produced a machine that could clean 50lb of cotton fibres a day. It consisted of a wooden cylinder on which were attached rows of slender spikes set $\frac{1}{2}$in apart. As the spikes revolved they meshed with a wire grid which formed part of the hopper that received the cotton seeds. The lint or fibres were pulled through the grid by the revolving spikes but the seeds were too large and were rejected. A patent was granted to Whitney on 14 March 1794 and he set up a business to manufacture cotton gins, as the machine became known, in partnership with a Phineas Miller at New Haven, Connecticut. The machine was essentially simple but its value was immediately appreciated,

and Whitney had to spend much time and money in litigation protecting his patent. The cotton gin was subject to a good deal of development in the early years of the nineteenth century and by the 1830s it was capable of dealing with 3cwt of cotton a day. One fundamental improvement was the substitition of a series of circular saws in place of the rotating spikes and in some quarters it was known as the saw gin. Some idea of its contribution to the growth of the American cotton industry can be assessed from the fact that in 1793, the year before Whitney's patent was granted, the United States exported 487,000lb of cotton while by 1803 this figure had risen to over 40 million lb.

Further Reading
Baines, E., *The History of the Cotton Manufacture in England*, 1835

47 POWER LOOM
John Austin (d 1820?)

Austin's power loom was the first of its kind to be built in Scotland, and its introduction led to the foundation of the modern Scottish textile industry.

John Austin was a native of Craigton near Glasgow. His first attempt to build a power loom was in 1789, only a few years after Cartwright had

Shuttle.

tackled the same problem in England (ref 45). His experiments did not reach a stage where a patent application would have been justified and he apparently abandoned the project. In 1796 his loom with certain improvements was demonstrated before the Glasgow Chamber of Commerce. Some sources suggest that it was due to the initiative of members of the chamber rather than to Austin's perseverance that the project was revived. Financial support was obviously given to Austin because in the next two years thirty looms were constructed and installed at Pollackshaws near Glasgow. These were apparently successful because the weaving shed was subsequently enlarged to accommodate a total of two hundred Austin looms, all of which were driven by a single steam engine. The loom was mainly of timber construction with a solid base and frame. Although it had a fairly complex mechanism, one weaver and a boy could look after between three and five looms.

Further Reading

Baines, E., *The History of the Cotton Manufacture in England*, 1835

SCIENTIFIC INSTRUMENTS

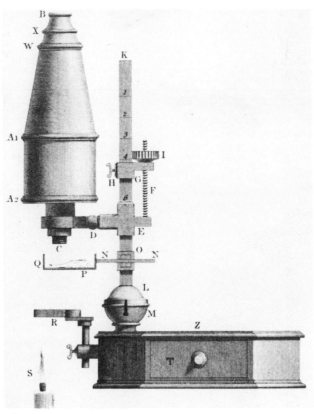

48 MICROSCOPE
John Marshall (1663–1725)

John Marshall was one of the earliest manufacturers of the compound microscope in Britain; his work formed a link between the seventeenth-century Italian pioneers such as Divini and Campani and the celebrated English school of instrument manufacturers which flourished in the mid-eighteenth century.

The earliest compound microscopes, in which two systems of lenses were used, were constructed in Italy in the middle of the seventeenth century. The tubes were invariably made of cardboard covered with decorated cloth, paper or leather, and the lenses for the eye-piece and object were encased in hardwood rings. As the instrument became more sophisticated, the method of focussing was improved and a number of accessories were added; the scientific instrument makers consequently became more dependent upon the skills of the wood turners, some of whom began to manufacture microscopes in their own right. John Marshall, who received his early training as a turner, was together with Edmund Culpepper the leading English manufacturer of microscopes in the early years of the eighteenth century. He made considerable improvements in the Italian sliding and screw barrel-type microscopes, and his later models showed at least a rudimentary ressemblance to the modern instrument. Like his Italian predecessors he was obliged to use cardboard for the tube, since thin brass sheet was not at that time generally available. He also used wood or sometimes ivory for the lens housings. The tube was supported by a brass rod attached to a rigid wooden base which was either circular or octagonal. Adjustment was by means of a coarse threaded screw on a lateral brass arm which permitted the instrument to be raised or lowered in relation to the object being studied. A ball and socket joint was also provided between the support rod and the base which allowed the instrument to be inclined to take advantage of the most favourable light.

Further Reading
Bedini, S. A. and De Solla Price, D. J., 'Instrumentation' *Technology in Western Civilisation*, New York, 1967

49 NAUTICAL QUADRANT
John Hadley (1682–1744)

During the early decades of the eighteenth century many instrument makers produced quadrants, but by 1740 it was generally agreed that John Hadley's reflecting quadrant surpassed all others. Although afterwards replaced by the sextant, it was an invaluable navigational aid and was widely used in ships of the two major maritime powers— Britain and France.

John Hadley's quadrant was one of several navigational instruments whose adoption contributed to the growth of British maritime supremacy during the eighteenth century. Hadley, who was the son of George Hadley, High Sheriff of Hertfordshire, first achieved fame for his telescopes, which were improvements on those of Newton and Gregory. He acquired great skill in the polishing of mirrors as a result of this work, and utilised this experience when he made the first reflecting quadrant in the summer of 1730. This was a considerable improvement over the earlier models and incorporated a

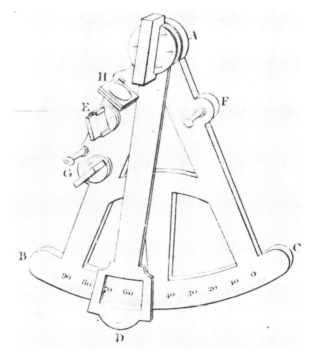

pair of mirrors located so that the observer could look straight at the horizon. In the zero position, the actual and reflected horizons were mutually aligned. The scale was graduated from 0° to 90°, although the arc subtended only 45° and the instrument should strictly have been called an octant. Hadley read a paper to the Royal Society on the 13 May 1731 entitled 'Description of a New Instrument for taking Angles', and in the following year tests were carried out by the Admiralty. In 1734 he made a further improvement by fixing

a spirit level so that a reading could be taken even if the horizon was not visible. A similar instrument was invented in America by Thomas Godfrey of Philadelphia a few months after Hadley completed his first quadrant.

Further Reading
Taylor, E. G. R., *The Mathematical Practioners of Hanoverian England 1714–1840*, Cambridge, 1966
Cotter, C. H., *A History of Nautical Astronomy*, 1968

50 REFLECTING CIRCLE
Tobias Mayer (1723–62)

Mayer's reflecting circle which was used for measuring lunar distances accurately was, like the quadrant, a major aid to navigation, and remained in service in improved versions for well over a century.

Tobias Mayer was a self-taught German mathematician who in 1752 was appointed Professor of Mathematics at the University of Göttingen. It was during his time at Göttingen that he published

his lunar tables which were based on the earlier theoretical work of the Swiss mathematician Euler. Many astronomers favoured the use of lunar tables as the most suitable method of determining longitude at sea, and Mayer devised his reflecting circle, or circle as it simply came to be known, in order to measure lunar distances accurately. In principle the instrument was similar to the quadrant. It consisted of a graduated circular limb with an index bar pivoted at the centre and was fitted with a Vernier scale at each end: as with the reflecting quadrant observations were taken with the aid of mirrors. Specula or glass mirrors were used at first but glass prisms were fitted on later models. The reflecting circle was highly regarded for its accuracy, and in some respects was considered superior to the sextant. Many navigators provided themselves with both circle and sextant, the former being used for lunar distances and the latter reserved for altitude observations. An improved version of the circle, known as the repeating circle, was in service through most of the nineteenth century, and it was only discarded when improvements to the sextant made it the highly regarded precision instrument it is today.

Further Reading
Cotter, C. H., *History of Nautical Astronomy*, 1968

51 IMPROVED MAGNETIC COMPASS
Godwin Knight (1713–72)

Knight's compass is generally regarded as the first to be constructed on scientific principles. It became the standard compass of the Royal Navy and together with the chronometer gave British seamen a decisive navigational advantage over their rivals.

After Godwin Knight had perfected his methods of making bar magnets (ref 98), he began experimenting to incorporate artificial magnets in the mariner's compass. In 1750 he read a paper to the Royal Society in which he discussed the faults of existing compass needles. He disclosed that he had examined needles from a number of leading makers and had found them all defective due to their shape which often meant that they had four poles instead of two. He recommended that a plain rhomboidal bar should be substituted and made a number of suggestions to improve needle suspension. John Smeaton, who had just set up in business as a scientific instrument maker, constructed a compass based on Knight's recommendations and is reported to have added some improvements of his own. The Admiralty showed interest and in 1751 Knight's compass was fitted in a number of vessels. Knight himself carried out a series of successful experiments on board HMS *Fortune* at Harwich in September of that year and in 1752 he was awarded £300 by the Admiralty—a paltry sum when compared with the chronometer prize of £20,000 offered by the Board of Longitude. The acceptance of the improved compass was nevertheless complete and it was fitted in all ships of the Royal Navy and certain large merchant vessels. Smeaton became engaged in other activities and the early Knight compasses were made by George Adams, an instrument maker in Fleet Street, with Knight signing his name on the card. In 1766 he took out a patent, no850, for further improvements mainly to dampen the effects of vibration. He also described in the same patent a reflecting azimuth compass similar to the type illustrated. This was an ordinary magnetic compass with an attachment which permitted observation of the sun's magnetic azimuth in order to discover the variation of the needle, or north-westing or north-easting as it was sometimes called. From this reading it was possible to correct the course. Knight spent the last years of his life as the principal librarian of the British Museum; he was appointed first holder of the post in 1756, at a salary of £160 a year, possibly as a recognition of his services to the nation.

Further Reading
Cotter, C. H., *A History of Nautical Astronomy*, 1968
Taylor, E. R. G., *The Mathematical Practitioners of Hanoverian England, 1714–1840*, Cambridge, 1966

52 ACHROMATIC LENS
John Dolland (1706–61)

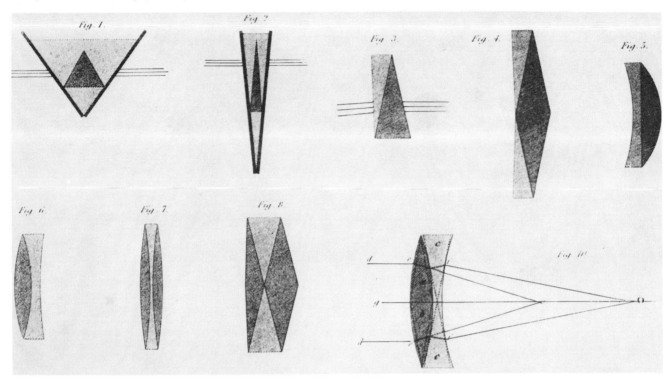

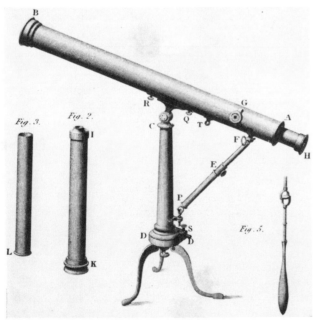

Dolland's achromatic lens was a major advance in optical and astronomical instrument manufacture. Thereafter colour aberrations were eliminated and image definition greatly improved.

The achromatic lens was one of several important inventions that became the subject of patent disputes during the eighteenth century. Credit for this major advance in optical instrument manufacture is almost universally given to John Dolland who was born at Spitalfields and trained originally as a silk weaver. It was not until fairly late in his life that he began to study optics after joining his son, Peter Dolland, who was already a practising optician. John Dolland began to carry out experiments relating to the controversy then raging over Newton's laws of refraction. By using prisms of water and glass to produce equal and opposite refractions, he found that the rays of light were parallel but strongly coloured. He continued to investigate the properties of various types of glass and eventually by joining wedges of crown and flint glass he was able to eliminate colour altogether. In 1758 he described his work in a paper entitled *Account of some Experiments Concerning the different Refrangibility of Light* which he read before the Royal Society. This work led to the achromatic telescope for which Dolland was awarded the Copley Medal of the Society. When John Dolland died the other London instrument makers thought they were free to use the achromatic lens without restriction but Peter Dolland claimed the patent as his own. In 1763 a group of his rivals brought a lawsuit against him and it was testified that the achromatic lens had originally been invented in 1733 by Chester More Hall. This was eventually allowed but, as Hall had never used his discovery commercially, the patent was not revoked.

53 COMPOUND MICROSCOPE
Benjamin Martin (1704–82)

Although a seventeenth-century invention, numerous improvements were made to the compound microscope by Benjamin Martin and other English instrument manufacturers, and by the end of the eighteenth century it had virtually assumed its established form.

Benjamin Martin, who contributed so much to the development of the microscope in the eighteenth century, was born at Worplesdon in Surrey and began life as a ploughboy. He was blessed with a natural mathematical talent which led him eventually to become a teacher of mathematics and an author of a number of scientific books. Martin set up in business as a scientific instrument maker in Fleet Street in 1750 but his first microscope is dated July 1738 which indicates that it was probably made in Chichester where Martin was earning his living as a schoolmaster. The instrument was described in a pamphlet published in that year and it appears to have been a rather crude form of compound microscope of the type originally developed in Italy during the latter half of the seventeenth century. Martin's early microscopes invariably had cardboard tubes and wooden lens holders but as he became established as an instrument manufacturer he began to work almost exclusively in brass with a consequent improvement in the accuracy and finish of the instrument. From around 1750 until the end of his life Martin produced a series of microscopes, each of a progressively higher standard. He described them collectively as 'universal' microscopes which meant that they could be used as either simple or compound instruments and that the body could be adjusted in an arc or radially by lengthening the transverse arm. In 1759 he introduced the 'between lens' mounted at the top of the snout which has been described as the only real improvement in the optical construction of the compound microscope during this period. Other features of Martin's work were his preference for the compass joint over the ball and socket joint favoured by many of the earlier manufacturers and his use of the tripod stand, both

Further Reading
Taylor, E. G. R., *The Mathematical Practitioners of Hanoverian England 1714–1840*, Cambridge, 1966

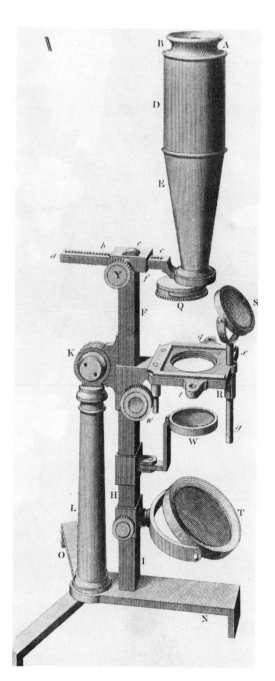

of which contributed to greater precision in focussing.

Further Reading
Clay, R. S. and Court, T. C. *The History of the Microscope*, 1932

54 MICROMETER
James Watt (1736-1819)

Watt's micrometer was the forerunner of the modern bench micrometer.

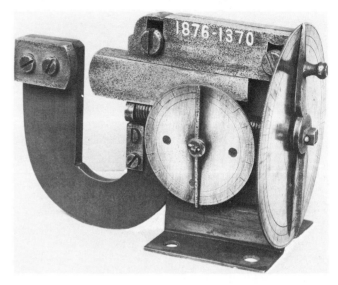

During the eighteenth century the accuracy of measuring instruments available to the engineer and instrument manufacturer was gradually improved. Verniers came into use in the middle of the century but their production was slow and expensive until Ramsden's linear ruling engine became available. Watt's micrometer was another major step forward. It was completed in 1772 before he came south to Birmingham to join Boulton. It comprised a U-shaped brass frame with a radiused iron anvil fixed to one limb. A second anvil was mounted on a slide on the other limb, and was moved by a screw engaging a rack cut on the lower edge of the slide. A dial was provided from which could be read off fractions of a revolution of the screw; the number of complete revolutions was registered by a pointer on a second separate dial. The screw had a pitch of 19tpi and the instrument could measure to an accuracy of 0.001in. Some years afterwards it was surpassed by a micrometer made by Henry Maudslay which had parallel anvils, a screw adjustment with a thread of 100tpi and could measure to an accuracy of 0.0001in.

Further Reading
Nickels, L. W., 'The Measurement of Length', *Engineering Heritage*, Vol 2, 1966

55 BAROMETER
Jean André Deluc (1727-1817)

In the hands of the master craftsmen of the eighteenth century the mercury barometer became a precision instrument, but its full potential was not realised until Deluc produced instruments based on his theories relating to barometric pressure.

The mercury barometer is a seventeenth century invention and is universally attributed to Evangelista Torricelli in either 1643 or 1644. Variations on Torricelli's original barometer soon appeared, and were usually classified according to the shape of the tube employed to contain the mercury. Jean Deluc, who was a Swiss geologist and physicist, spent many years investigating the problems relating to the accurate measurement of barometric pressure and published his findings in 1772 in his classic work *Recherches sur les Modifications de l'Atmosphère*. The most important consequence of Deluc's research was that it became possible to measure height accurately by means of a barometer. Deluc, in fact, by discovering the law linking air pressure, altitude and temperature, invented the altimeter or made that instrument possible. Deluc was a member of the international fraternity of science which emerged in the second half of the eighteenth century. He became a Fellow of the Royal Society and from 1773 to the end of his life he was domiciled permanently in England. He made a number of scientific discoveries and, as well as his improved barometers, he invented a whalebone hygrometer which recorded the moisture content of the atmosphere on a large dial. His other work related to latent heat, where he covered much of the same ground as Joseph Black, and to the study of vapour pressures. It was Deluc who originated the theory that the quantity of water vapour that can be contained within a space is independent of the pressure of any other gas in the space, and in this research he anticipated John Dalton by many years. When he arrived in England, Deluc was appointed Reader to Queen

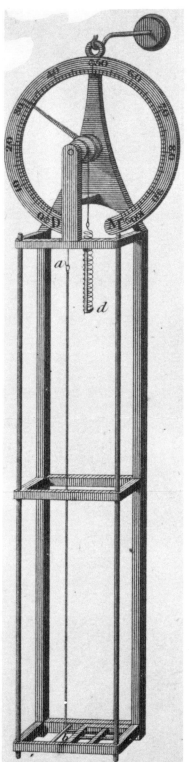

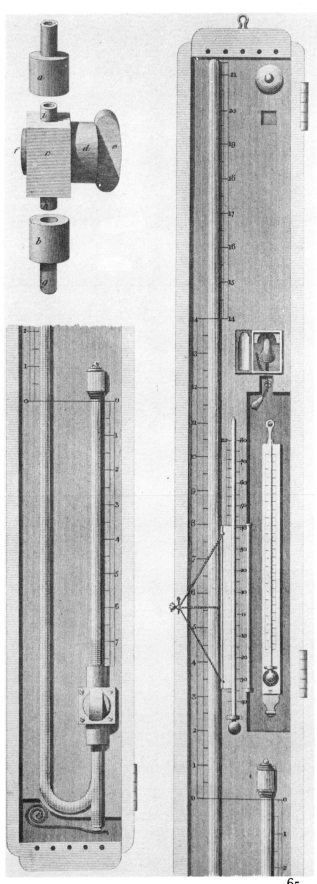

Charlotte, a post that provided him with a satisfactory income and enabled him to pursue his scientific researches. He died at Windsor in 1817.

Further Reading
Goodison, N., *English Barometers 1680–1860*, 1969

56 DIVIDING ENGINE
Jesse Ramsden (1735-1800)

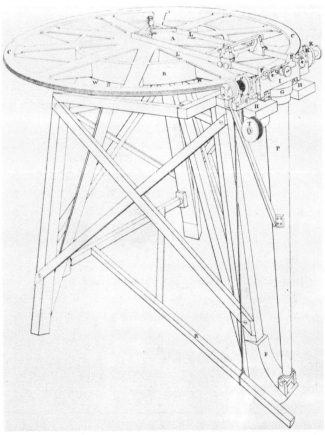

The high standard of accuracy attained by mathematical and astronomical instrument makers in Britain during the latter part of the eighteenth century would not have been possible without the dividing engine of Jesse Ramsden.

Born in Halifax, Ramsden was apprenticed as a cloth maker but moved to London in 1755. In 1758 he became an apprentice to a mathematical instrument maker and eventually worked with many of the leading craftsmen of the day, including John Dolland whose daughter he married in 1765 or possibly a year later. In 1763 Ramsden set up in business on his own in the Haymarket and began to consider the problems relating to the construction of a dividing engine. Fine screw threads of the greatest accuracy were required and Ramsden first had to make a series of screw cutting lathes; the products of one machine were incorporated into the next to achieve a progressively higher standard. His first dividing engine was described in a pamphlet issued in 1771 which included a preface by Nevil Maskelyne, the Astronomer Royal. In 1773 he received an award of £615 from the Board of Longitude for his second dividing engine on the condition that he published a description of it. This was made after a sextant scale divided by the machine had been approved by John Bird who had hitherto been the acknowledged master of scale division by geometrical methods. Ramsden also invented or improved various other scientific and mathematical instruments including a theodolite, micrometer, barometer and an assay balance, and he carried out experiments to determine the specific gravity of fluids for which he was awarded the Copley Medal of the Royal Society. His dividing engines for circular and linear scales, however, were acknowledged as his principal contribution to the science of instrument manufacture. An attempt was made to prevent models from being sold abroad and it is said that the first Ramsden engine introduced into France was concealed in the pedestal of a table.

Further Reading
Taylor, E. G. R., *The Mathematical Practitioners of Hanoverian England, 1714-1840*, Cambridge, 1966
Rolt, L. T. C., *Tools for the Job*, 1965

57 REFLECTING TELESCOPE
William Herschel (1738-1822)

Herschel's giant reflecting telescope opened up a new era in optical astronomy; this was due principally to the high quality of the specula or mirrors which Herschel was able to produce.

William Herschel was one of several notable scientists in the eighteenth century who made a lasting reputation in one profession after working assiduously for some years in a wholly different field. Herschel was the son of a Hanoverian bandmaster who served in England and was himself trained as an army musician. He left the army in his early twenties and served successively as a church organist in Halifax and afterwards at the Octagon

Chapel in Bath. His interest in science gradually took up more and more of his time, and in 1773 he began to construct a series of telescopes which became progressively more powerful. In 1780 he discovered the planet Uranus, and this fact, plus possibly his Hanoverian connections, brought him to the notice of George III. In 1785 the king arranged for Herschel to receive a grant of £2,000 to construct a large reflecting telescope. In April of the following year Herschel moved to a house at Slough and began work on the giant reflector. The major problem was the casting, grinding and polishing of the large specula which served as mirrors in eighteenth-century telescopes. These were essentially discs cast from a copper-tin alloy with a silvery appearance which could be polished to give a high degree of reflectivity. The first great speculum was inserted in the tube in 1787 but the second cracked on cooling and a third had to be made. George III followed the progress closely and arranged for a further grant to be made with an allowance for repairs of £200 per annum. When completed the telescope easily dwarfed all of its predecessors, the tube being 39ft 4in long and with a diameter of 49½in. An inclination of about 3° caused the mirror to throw an image slightly to one side of the tube where the eye-piece was located. The observer stood with his back to the sky and gave instructions to assistants below through speaking-tubes. The tube was arranged to swing round on a circular track and elevation was controlled through a system of multiple pulleys. The telescope immediately became an object of great interest and scientists and laymen flocked to see

it from all over the world. Herschel was universally honoured and was created first President of the Royal Astronomical Society, a fitting reward for his work in optics which had truly opened up a new era in astronomy. It was said that before he began it was almost unknown for a star to be seen without distortion, apparently owing to rays or 'tails' of light. Herschel was the first to see stars as round objects in the sky—to use his own words—'as round as a button'.

Further Reading
King, H. C., *History of the Telescope*, 1955

58 IMPROVEMENTS IN THE MANUFACTURE OF ASTRONOMICAL INSTRUMENTS
Edward Troughton (1753–1835)

Troughton's only patent referred to instrument manufacture in the broadest terms, but the improvements which he introduced were of the utmost significance, and profoundly influenced the subsequent evolution of astronomical and nautical instruments.

Edward Troughton was born in Cumberland, but moved to London and became an apprentice instrument maker in the firm of his brother John Troughton. He suffered from the family defect of colour-blindness, which prevented him from undertaking optical work, but he eventually became a partner and took over control of the company on his brother's death. His ability as a craftsman was second to none, and throughout his long working life he was responsible for a number of inventions which he never attempted to patent. In 1778 he introduced a method of graduating arcs of circles which was afterwards described by Sir George Airy as the greatest single improvement ever made in the

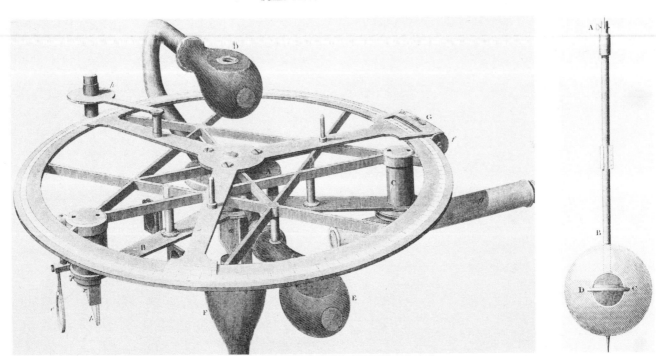

art of instrument making. The first instrument incorporating this invention was an astronomical quadrant of 2ft radius which he made in 1785. Three years later he was granted patent no1644 for 'a method of framing to be used in the construction of octants, sextants and other nautical and astronomical instruments whose limbs are formed on a circle or part of a circle'. Astronomical instruments made by Troughton were installed at Greenwich and other major observatories all over the world. He also manufactured theodolites of great precision and improved marine barometers, pyrometers and other instruments. He developed two different types of compensated pendulum, one of which was a mercury pendulum that was more simple and compact than Graham's earlier version. He was also the first to substitute fragments of spider's web in place of metal wires in view finding.

Further Reading

Taylor, E. G. R., *The Mathematical Practitioners of Hanoverian England, 1714–1840*, Cambridge, 1966

59 THEODOLITE
Jesse Ramsden (1735–1800)

The two great theodolites designed and manufactured by Jesse Ramsden were used in the triangulation exercise to connect the surveys in England with those in France, and also in the subsequent triangulation of Great Britain and Ireland.

Interest in topographical surveying in Britain during the second half of the eighteenth century was stimulated by military requirements, and the origins of the ordnance survey can be traced to the attempt to map the Highlands in 1747 after the Battle of Culloden. Other programmes were undertaken spasmodically but were interrupted by the Seven Years' War and the War of American Independence. In 1783 an approach came from France suggesting that it would be desirable to construct a series of triangles from the vicinity of London to Dover and from there to connect the triangles with those already executed in France. Jesse Ramsden, as the leading mathematical instrument maker of the day, was instructed to produce a number of instruments for the use of the surveying team, who were under the direction of Major-General William Roy. The first stage of the operation, the accurate measurement of the base line on Hounslow Heath, was completed in 1784, but progress was then delayed for three years until Ramsden had finished his first great theodolite which was intended to measure the angles in the triangulation work. A primitive form of theodolite had existed as early

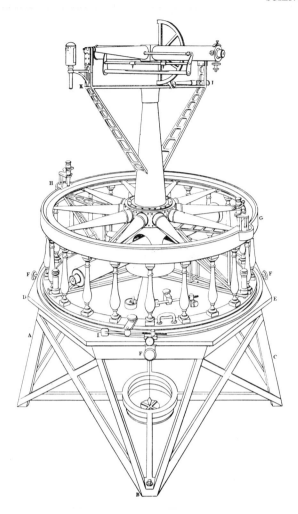

as the sixteenth century, and the instrument had already been progressively improved in the eighteenth century by Sissons and, in particular, by the addition of a refracting telescope embodying the achromatic lens of John Dolland (ref 52). Ramsden's instrument, however, was to set a new standard of accuracy, being constructed with the aid of his own dividing engine (ref 56) and it remained in constant use until 1862. It had a 3ft horizontal brass circle divided to 10' and made to read to 1″ by means of two micrometer microscopes. The circle was attached to a central axis on which was mounted the telescope; and the telescope axis and supporting bar could be aligned horizontally with the aid of two spirit levels. When it was used for the triangulation of Great Britain and Ireland, eight sights were taken over a distance of 100 miles, and it was estimated that the probable error for a distance of 70 miles was only 5″ Ramsden's first theodolite was used for the first time at the triangulation station set up at the Hampton Poorhouse on 31 July 1787. His second instrument, which was also used on the Anglo-French survey and afterwards in the British Isles, was completed in 1790.

Further Reading
Richeson, A. W., *English Land Measuring to 1800*, Cambridge, Mass, 1966

60 MAXIMUM AND MINIMUM THERMOMETER
James Six (d 1796)

The maximum and minimum thermometer became one of the most familiar and useful instruments in meteorology and in many other branches of science.

Numerous thermometers were invented during the eighteenth century and there were upwards of a dozen different temperature scales in use in the same period. These included three which have survived to the present day, namely those of Réaumur, Celsius and Fahrenheit. One of the most important developments in the science of temperature measurement occurred in the last decade of the century with the invention of the maximum and minimum thermometer by James Six, a mathematical instrument maker of Maidstone. Six described his instrument in a pamphlet published in 1794 entitled *Construction and Use of a Thermometer*

for showing the Extremes of Temperature in the Atmosphere during the Observer's Absence. It consisted of a continuous glass tube shaped in the form of a double U. The thermometric fluid in the centre leg was alcohol and a further quantity of alcohol was included at the other end of the tube. In between these two volumes of alcohol was a quantity of mercury which merely served to move the indices; the latter were made of steel wire and covered with glass. As the alcohol expanded or contracted depending on the variation in temperature, the mercury moved around the tube and pushed the indices before it. Friction was sufficient to hold the indicator in position to record the maximum or minimum temperatures. Mixed liquid thermometers, such as the type devised by Six, had an inherent defect as eventually the alcohol would 'wet' the glass and penetrate between the mercury and the glass. Later in the nineteenth century all-mercury thermometers of this type were introduced.

Further Reading

Taylor, E. R. G., *The Mathematical Practitioners of Hanoverian England, 1714–1840*, Cambridge, 1966

Middleton, W. E. K., *A History of the Thermometer*, Baltimore, 1966

CLOCKS
AND WATCHES

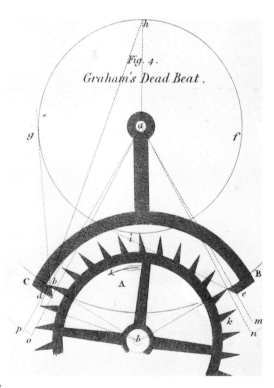

Fig. 4.
Graham's Dead Beat.

61 THE DEAD-BEAT ESCAPEMENT
George Graham (1673-1751)

Graham's dead-beat escapement became the standard observatory clock mechanism throughout the world for nearly two hundred years and, together with his mercury pendulum (ref 62), was a major contribution to horological science.

The common escapement fitted to clocks in the early years of the eighteenth century was the anchor escapement, introduced by Clement. This was superseded by George Graham's dead-beat escapement which he invented in about 1715. Graham's device eliminated recoil and offered less restriction to the pendulum; it was a major advance in clock-making and, when used in con-junction with the mercury pendulum or the grid-iron pendulum of John and James Harrison, it was regarded as the standard mechanism for observatory timekeepers until the end of the nineteenth century. Graham also invented a cylinder escapement for watches in 1726 which greatly improved the degree of accuracy in timekeeping, a discrepancy of no more than a minute a day being obtainable with average fluctuations in temperature.

Further Reading
Lloyd, H. A., *Some Outstanding Clocks over Seven Hundred Years 1250–1950*, 1958

62 COMPENSATED PENDULUM
George Graham (1673-1751)

Graham's compensated pendulum was one of the most important advances made in eighteenth-century horology and resulted in a much higher standard of accuracy in timekeeping.

George Graham was born at Kirklinton in Cumberland and followed in the footsteps of his illustrious uncle Thomas Tompion when he became a watchmaker's apprentice in London. He eventually succeeded to his uncle's business in Water Lane, off Fleet Street, and was recognised as the most accomplished craftsman of his day. A number of clockmakers in the early part of the eighteenth century were concerned with the inaccuracy in timekeeping caused by effects of temperature on the length of a clock's pendulum. As early as 1715, Graham was considering a compound or grid pendulum using iron and brass, but it was not until 1726 that he produced his compensated mercury pendulum, which he described in a paper delivered before the Royal Society with the rather ponderous title of *A Contrivance to avoid the Irregularities in a Clock's Motion occasioned by the Action of Heat and Cold on the Pendulum Rod*. The mercury pendulum was simple and effective. When the rod, which could be made either of brass or steel, was subjected to a rise in temperature it expanded, thereby increasing the period of swing. This was countered by a hollow bob at the end of the rod which was

partly filled with mercury. The large coefficient of expansion of the liquid metal caused it to expand upwards, thus compensating for the linear expansion of the rod, and the position of the centre of gravity of the bob remained constant. The Harrisons, James and John (ref 63), also produced a compensated pendulum about this time, and in Italy Bernado Facini of Venice made a four-day clock with a compensated pendulum comprising silver and steel components.

Further Reading

Lloyd, H. A., *Some Outstanding Clocks over Seven Hundred Years 1250–1950*, 1958

Taylor, E. R. G., *The Mathematical Practitioners of Hanoverian England, 1714–1840*, Cambridge, 1966

63 CHRONOMETER
John Harrison (1693–1776)

Harrison manufactured the first chronometer and, although others who followed produced timepieces that were easier to manufacture in quantity, it was his work which gave the English navigators from Cook onwards an inestimable advantage over their French and Spanish rivals.

John Harrison was the son of a Yorkshire carpenter. His education was scanty, but he showed a natural mechanical aptitude and joined his father in his workshop at an early age. In 1715 he constructed an eight-day clock with a wheel train made entirely of wood; the clock worked for over a hundred years. His next important clock was one fitted with a grid iron pendulum similar to the type introduced by George Graham (ref 62). Harrison moved to London, and was at once recognised as an horologist of the first rank. He decided to compete for the prize offered by the Board of Longitude which had been constituted in 1713 by Act of Parliament, the purpose being to stimulate or devise a method of finding the longitude at sea. The board had laid down that the position must be determined within sixty, forty or thirty

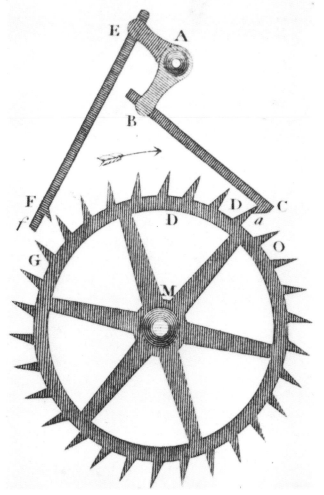

geographical miles and offered rewards of £10,000, £15,000 and £20,000. To achieve this a timekeeper of the highest accuracy was required. Harrison worked patiently on the project and completed his first instrument in 1735, for which he received a certificate from the Royal Society to say that he deserved encouragement. This first chronometer was taken by Harrison in the following year on a voyage to Lisbon and back in HMS *Centurion*. It performed well and enabled Harrison to correct an error in the reckoning, but it was not of sufficiently high standard to merit one of the major prizes. Harrison accordingly made a second chronometer which was completed in 1739 and was less cumbersome than the first. His third chronometer, for which he was awarded the Copley Medal by the Royal Society, was even smaller and for this he received a grant from the Board of Longitude to enable him to continue further work. His perseverance was rewarded, and when he completed his fourth chronometer in 1759, which is illustrated above, the board was sufficiently impressed to order further extensive trials at sea. William

Harrison, his son, took the chronometer to Jamaica and back between 18 November 1761 and 26 March 1762 in HMS *Deptford*, during which time it only lost 1min 54sec—an error of eighteen geographical miles. Further trials were held in 1763-4 in which Nevil Maskelyne the Astronomer-Royal participated and, after some deliberation, the board awarded Harrison £10,000 in 1765. He disputed the sum, and in order to secure redress he made a fifth chronometer at the age of seventy-nine which was placed in the private observatory of George III at Richmond. With the king's support he petitioned parliament and was eventually given the balance of the award due to him. The accuracy of the Harrison chronometer was due to his use of a bi-metallic 'curb' to vary the effective length of the balance spring.

Further Reading
Taylor, E. R. G., *The Mathematical Practitioners of Hanoverian England, 1714-1840*, Cambridge, 1966
Lloyd, H. A., *Some Outstanding Clocks over Seven Hundred Years 1250-1950*, 1958

64 CHRONOMETER
Pierre Le Roy (1717-85)

Le Roy's chronometer was not quite as accurate as that of John Harrison, but it employed a method of correcting temperature errors that was later adopted for close time keeping to the exclusion of all others.

Pierre Le Roy was the eldest son of Julian Le Roy, who was himself one of the most gifted of the French horological inventors in the eighteenth century. Pierre followed his father's profession and is generally considered to have surpassed the achievements of his father, chiefly on account of his contributions to the development of the marine chronometer. France had no precise counterpart to the Board of Longitude in Britain but, as a leading maritime nation, the French were equally aware of the need for exact timekeeping at sea. In 1763 the Academy of Sciences offered a prize for a chronometer, for which the two chief contenders were Pierre Le Roy and Ferdinand Berthoud. Le Roy entered two chronometers and was judged to be the winner. His chronometers did not perform as well as those of John Harrison for, when tested over a period of forty-six days during a voyage in

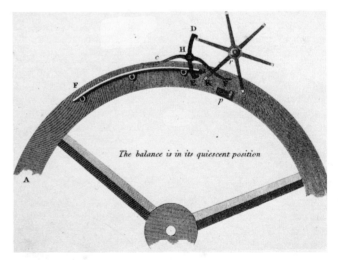

The balance is in its quiescent position

the Channel and North Sea, they lost seven and thirty-eight minutes respectively. Nevertheless when considered by any less exacting standards they must be regarded as instruments of a very high degree of precision. Le Roy adopted a form of escapement which in principle was similar to that devised by Mudge in England. This had the merit of ensuring freedom of the balance from the influence of the train after the impulse had been delivered. His chief contribution to the development of the marine chronometer was his temperature compensation device which, if not superior to Harrison's, could be reproduced more cheaply. The device was mounted on the balance wheel and had the effect of varying the radius of gyration of the wheel with a change of temperature. Thus, with a fall in temperature, the inertia of the balance was increased, while with a rise it was correspondingly reduced. The French government rewarded Le Roy with a pension of 1,200 livres for his work, and on his death in 1785 continued to pay half of that sum to his widow.

Further Reading
Daniels, G., *English and American Watches*, 1967
McCloy, S. T., *French Inventions of the Eighteenth Century*, Kentucky, 1952

65 CHRONOMETER
John Arnold (1736–99)

The final development of the marine chronometer took place during the last twenty years of the eighteenth century, and the pre-eminence of the English watchmakers such as John Arnold in this branch of horology gave the Royal Navy an immeasurable advantage over its continental rivals.

Harrison's chronometer (ref 63) was an instrument of great precision but it was difficult and expensive to reproduce in quantity. It was left to John Arnold and his contemporary Thomas Earnshaw (ref 67) to produce chronometers in large numbers and of sufficiently high standard to satisfy the conditions imposed by the Board of Longitude. Arnold was born in Bodmin and was apprenticed to his father who was a watchmaker in the town. After working for a time in Holland, he eventually set up in business in London at Devereux Court, in the Strand. After a few years, when his business began to prosper, he moved to the Adelphi; it was about this time that he was introduced at Court and made several timepieces specially for George III. One of these was a watch a little over half an inch in diameter with a ruby cylinder escapement and repeating mechanism. In 1770 he made his first marine chronometers but their performance was disappointing when tried at sea. He worked patiently on the problem for several years and in 1775 he patented the helical balance spring and a balance containing a device for temperature compensation which was a preferable alternative to Harrison's compensation curb. Arnold's improvements were not only technically sound but they simplified construction and he was able to produce chronometers that were small enough to be carried in the pocket. The performance of some of Arnold's chronometers has never been surpassed; one which

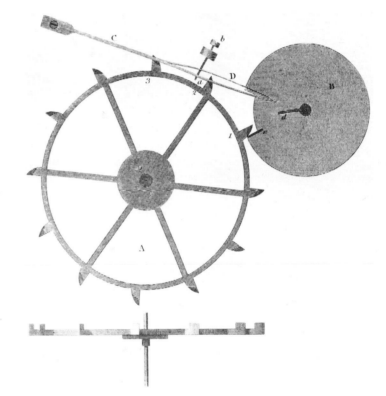

he sent to Greenwich for trial was tested for a year after which it was found that its greatest daily error was only three seconds. Various small sums of money were paid out to Arnold during his lifetime for his work on chronometers by the Board of Longitude, and after his death his contribution was acknowledged belatedly by the board when they made a further grant to his son, John Roger Arnold, who continued to make chronometers in large numbers for the Admiralty and the East India Company.

Further Reading
Daniels, G., *English and American Watches*, 1967

75

66 DETACHED LEVER ESCAPEMENT
Thomas Mudge (1717–94)

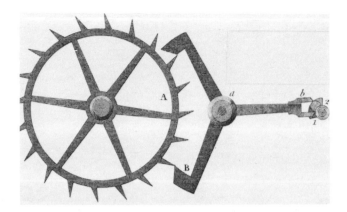

The watches made by Thomas Mudge were timekeepers of the highest degree of precision, but they were too delicate and expensive to be generally practical. However they incorporated a form of escapement that was the antecedent of all escapements used in good quality watches today.

Mudge was an apprentice of George Graham and his work, like that of his master, displayed a standard of craftsmanship of the highest order. He was a Westcountryman by birth, and received his education at Bideford Grammar School where his father was a master. After his period of apprenticeship and service as a journeyman in Graham's employment, he set up in business as a clockmaker in partnership with William Dutton at 77, Fleet Street. His principal contribution to horological science was his detached lever escapement, first incorporated in a watch made for George III which he subsequently gave to Queen Caroline. The principle which governed the lever escapement was the need to free the balance from all extraneous interferences. This he achieved by unshackling the balance from the lever and terminating the latter in a two-pronged fork. Mudge devoted the second half of his working life, like many of his contemporaries, to the problem of producing an accurate marine chronometer. In 1776 he was appointed the king's watchmaker, and in the same year he completed his first chronometer for which he was awarded £500 by the Board of Longitude. He made two more chronometers which were subjected to rigorous tests by Nevil Maskelyne, the Astronomer-Royal, who rejected them on the grounds that they did not satisfy the board's requirements. A long argument ensued in which both sides published pamphlets to support their case, and eventually in 1792 Mudge was awarded £2,500 by a committee appointed by the House of Commons. The watches and chronometers produced by Mudge were afterwards copied extensively in the nineteenth century. The high standards which he set were followed by his successors, and the good quality watches that became available at moderate prices incorporated many features of his design.

Further Reading
Daniels, G., *English and American Watches*, 1967
Lloyd, H. A., *Some Outstanding Clocks over Seven Hundred Years 1250–1950*, 1958

67 CHRONOMETER
Thomas Earnshaw (1749–1829)

Thomas Earnshaw shared with his contemporary and rival John Arnold (ref 65) credit for the final development of the marine chronometer.

Watch and clock making in England continued to flourish throughout the eighteenth century, and the high standards set by Graham, Mudge and Harrison were maintained during the later decades by master craftsmen such as Arnold and Earnshaw. Thomas Earnshaw was a native of Ashton-under-Lyne who became a journeyman watchmaker in the employment of John Brockbank. He eventually set up in business on his own at 119, High Holborn, and one of his early tasks involved repairing Graham's transit clock at Greenwich. In 1780 at the age of thirty-one Earnshaw evolved an escapement similar in principle to Arnold's which was applied eventually to a number of marine chronometers. Some authorities consider that it was superior to Arnold's escapement, and the pocket watches in which it was incorporated were highly regarded. The Board of Longitude gave an impartial judgment when they awarded Earnshaw the sum of £3,000 for his improvements to the marine chronometer;

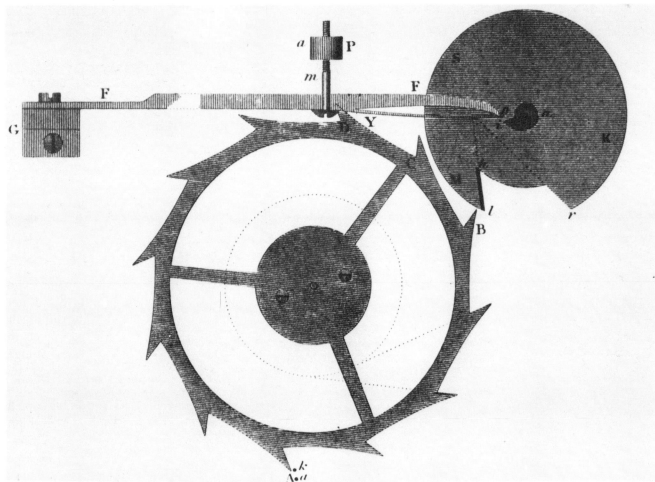

this was identical to the sum given to John Roger Arnold in relation to his father's work in 1799. Earnshaw was supported in his claim by Dr Maskelyne, the Astronomer-Royal, but it was opposed by Sir Joseph Banks who, as President of the Royal Society, was an ex-officio member of the Board of Longitude. The grant was eventually made in 1805.

Further Reading

Daniels, G., *English and American Watches*, 1967

Taylor, E. G. R., *The Mathematical Practitioners of Hanoverian England, 1714–1840*, Cambridge, 1966

MACHINE TOOLS

68 ROLLING MILL
Kristofer Polhem (1661–1751)

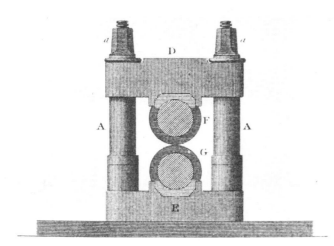 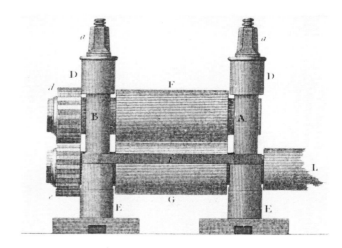

The introduction of power-driven rolls increased the output of metal plates and resulted in a higher quality product; one of the consequences was an adequate supply of iron and copper plates for steam engine boilers which were required in increasing numbers as the Industrial Revolution gained momentum.

Kristofer Polhem, the Swedish scholar, engineer and industrialist, is generally credited with the invention of power-driven plate rolling machinery in about 1734. In the first half of the eighteenth century Swedish iron was considered the best of any manufactured in Europe and the Swedish iron-masters, who were jealous of their profitable export markets, endeavoured to keep to the forefront of technical advances in iron manufacture. Polhem's rolls were invariably driven by water power which was abundant in Sweden. They comprised an upper and lower roll connected by a simple gear train. The rolls themselves were forged from iron and then turned on a lathe which would appear to be

the forerunner of the heavy industrial lathes used in Britain at the end of the century. Finally the rolls were hardened and polished to impart the optimum finish to the surface of the metal plates. At least two men were required for the rolling operation, one on each side of the rolls. The iron billet was heated in a furnace and fed in between the rolls by one or more operators with the aid of tongs. It was received at the other side and returned so that the operation could be repeated until a plate of the desired uniform thickness was obtained. In Britain the power-driven rolling mill was not introduced as early as in Sweden, but after the invention of Henry Cort's puddling process (ref 11) it became rapidly established and steam power as well as water power provided the motive force.

Further Reading
Hulton, K. G. P., *The Machine*, New York, 1968
Rolt, L. T. C., *Tools for the Job*, 1965

69 GEAR CUTTING MACHINE
Henry Hindley (1710–71)

Hindley's gear cutting machine was intended only for clock gears, but it was a notable advance on anything previously produced and incorporated a system of differential indexing a hundred and fifty years before the same feature appeared on the milling machine.

The high standards of craftsmanship achieved in

clock manufacture during the eighteenth century would not have been possible without a corresponding improvement in clockmakers' tools. One significant advance was the development of the gear cutting machine and its gradual evolution into an industrial machine tool. Henry Hindley of York, who was responsible for a cutting engine featuring

before the latter moved to London, and it was Smeaton who left the most complete description of Hindley's machine in a paper presented to the Royal Society, although this was not delivered until about forty years after he had seen it in Hindley's workshop in about 1741. The machine had a sturdy box frame construction which gave it greater rigidity than any of its predecessors. Accurate adjustment of the depth of tooth could be achieved through a lead screw operated from one side, while another handle controlled an index plate with the circles divided for all common numbers of teeth used in clock trains, from 49 to 365. Among the

advantages of Hindley's machine was the facility with which its cutter could be removed for re-sharpening, and the teeth being cut with flat root spaces instead of concave surfaces, as had been the case with earlier engines. This was achieved by means of the large handle and yoke which enabled the cutter to be moved parallel to the axis of the gear. A special attachment was also provided to cut racks and pinions.

Further Reading

Woodbury, R. S., *History of the Gear Cutting Machine*, Cambridge, Mass, 1958

70 BORING MACHINES
John Wilkinson (1728–1808)

The cannon boring machine patented by John Wilkinson in 1774 was similar to that installed four years earlier at Woolwich Arsenal and the patent was subsequently revoked. His second machine was designed specifically for cylinder

boring, and it was used for the cylinders of the first two commercially-built Watt steam engines. It is rightly regarded as a landmark in the development of machine tools.

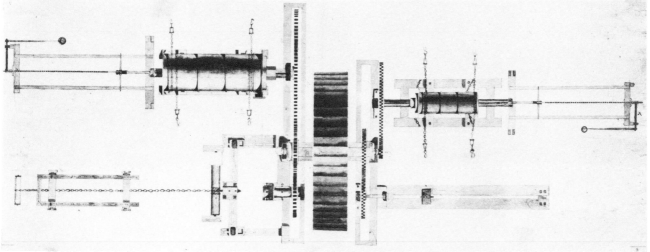

John Wilkinson, the most celebrated of eighteenth-century ironmasters, was born of yeoman stock at Clifton in Cumberland. His father, Isaac Wilkinson, was originally a farmer but later went into business as an iron smelter and manufacturer of a laundry box iron which he patented in 1738. The younger Wilkinson also became an iron smelter, and eventually, became manager and owner of the works that his father had established at Bersham near Wrexham. John and his brother William, who was an equally forceful character, continued to conduct the business after their father's retirement, and John, shortly after his second marriage in 1763, took over a blast furnace and colliery at Willey, near Brosley in Shropshire. It was at Willey that Wilkinson experimented with a cannon boring machine for which he was granted patent no1063 on 27 January 1774. In this machine the cannon, which had been cast solid, was rotated in a horizontal plane while the boring head and bar were fed in progressively by means of a rack and pinion. A large

handwheel was provided for the operator to control the rate of feed. This machine was similar to that installed in Woolwich Arsenal in 1771, and the Board of Ordnance eventually challenged Wilkinson's patent, which was revoked. Long before this occurred Wilkinson's cannon-boring machine had achieved a place in history by successfully boring the cylinder for the experimental Watt steam engine known as the Kinneil engine—a feat which had been beyond the capabilities of the Smeaton boring mill at the Carron works in Scotland. Wilkinson perceived that his relatively small cannon boring machine would be unsuitable for the cylinders of the larger steam engines which Boulton, Watt & Company intended to build, and accordingly he constructed a second machine designed specifically for dealing with steam engine cylinders. Details of this machine are shown in the second illustration; the cylinder was fixed upon a work table and the revolving boring bar passed through the inside. There can be no doubt that the machine was an outstanding success,

81

although Wilkinson inexplicably failed to take out a separate patent on it. He continued to supply cylinders to Watt at the Soho works, but similar machines were also built by his competitors after 1779 when his first patent was revoked. Wilkinson continued to conduct his iron smelting business successfully, although he eventually parted company from his brother after a long drawn out dispute. In July 1787 he launched the first of a number of iron barges to carry castings down the Severn from his Coalbrookdale works; this was the first recorded use of iron as a constructional material in ship building. He also patented a machine for making lead pipes (ref 73) and introduced many innovations in the process of iron smelting.

Further Reading
Steeds, W., *A History of Machine Tools 1700–1790*, Oxford, 1969
Hoyland, J., *Engineering Heritage*, Vol 2, 1966
Rolt, L. C. T., *Tools for the Job*, 1965

71 LATHE
Jacques Vaucanson (1709–82)

Vaucanson's lathe, which was made about twenty years before Henry Maudsley's first lathe, represents the transition between the wooden-bed ornamental lathe and the all-metal precision engineering lathe.

Jacques Vaucanson was a man of many talents who made a number of notable contributions to engineering technology in France during the eighteenth century. His lathe, which was built within the period 1770–80, now rests in the Conservatoire National des Arts et Métiers, and is the oldest engineering lathe in existence. The lathe was unconventional in many respects, being contained within a framework of square-section iron bars which were bolted together. The bed consisted of two additional iron bars measuring about $1\frac{1}{2}$in square which were set on edge so that the carriage could slide on two faces, each inclined at an angle of 45° to the horizontal. This ensured a much greater degree of precision than had hitherto been obtained by his predecessors. The saddle was a brass casting and was provided with a screw-operated cross-slide. A square-thread screw was fitted which permitted the saddle to traverse the length of the bed-ways, and this again was an advance over earlier lathes. The work-piece was supported between centres, but only a limited longitudinal adjustment was provided, and this has led to speculation that the lathe may have been designed to machine one particular component. A further mystery concerns the method of drive since there was no provision for a pulley system, one theory being that the pulley was attached in some way to the workpiece and driven from an external wheel.

Further Reading
Rolt, L. T. C., *Tools for the Job*, 1965
Steeds, W., *A History of Machine Tools*, Oxford, 1969

72 GEAR CUTTING MACHINE
Samuel Rehé (d 1806?)

Samuel Rehé's gear cutting machine marked the transition between the clockmaker's tool and a machine capable of producing gear wheels for a variety of general engineering applications; it can be regarded as the forerunner of the modern gear cutting machine.

The gear wheel cutting engine of Samuel Rehé was more robust than that of Henry Hindley, being equipped with a massive cast iron frame to give the necessary rigidity for cutting larger gear wheels.

Rehé was a mathematical instrument maker of Shoe Lane in London and he probably intended that his machine should be used for other applications than simple clock gear wheels. The index plate had a diameter of 19in and was divided by Ramsden's dividing engine (ref 56). Rehé attached great importance to the design of the cutting tools which show a close similarity to a modern milling cutter. In some cases the cutters were formed from a single bar of steel but inserts were also used. To ensure that the cutters would give a good performance, Rehé also produced a special grinding machine which is shown in the smaller illustration. This maintained a sharp cutting edge on each of the teeth and an attachment was provided to shape the profiles of the cutters. The gear cutting machine was also fitted with attachments for cutting worm wheels and racks, and also for cutting annular wheels which made it considerably more versatile than its predecessors.

Further Reading
Woodbury, R. S., *History of the Gear Cutting Machine*, Cambridge, Mass, 1964

73 LEAD PIPE DRAWING MACHINE
John Wilkinson (1728–1808)

John Wilkinson's pipe drawing machine produced a superior quality lead pipe and remained in common use for over a hundred years. The demand for lead pipe, particularly for domestic plumbing, rose rapidly during the Industrial Revolution, and Wilkinson's machine enabled manufacturers to meet all requirements.

John Wilkinson, as one of the first industrialists,

had numerous other business interests in addition to iron smelting and boring cylinders for Boulton & Watt engines. He was closely involved in the lead industry, owning shares in lead mines, and was actively concerned with the problems of lead smelting and manufacture. In particular he studied the current method of making lead pipes. This involved casting the lead into a circular mould

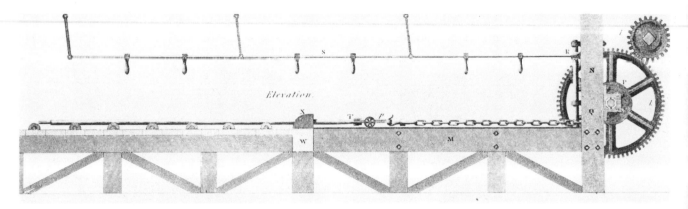

Elevation.

round a mandrel or core bar. Local defects were common, and the technique was generally unsatisfactory especially from the manufacturer's point of view, as the walls of the pipe were often unnecessarily thick. Wilkinson applied himself to the problem with his customary diligence, and in 1790 took out patent no1735 for a pipe drawing machine. The earlier casting sequence was retained, but the wall thickness of the pipe was reduced by drawing it through a die on a bench, similar to the type employed in wire production. The reduced pipe

was of greater length and more uniform in quality. After his boring mill (ref 70), the pipe drawing machine was probably Wilkinson's most successful invention. He used it himself on a large scale, and it proved very profitable to him; when the patent expired it came into general service, until it was eventually replaced by the extrusion press.

Further Reading

Hoyland, J., 'John Wilkinson—Man of Iron', *Engineering Heritage*, Vol 2, 1966

74 WOODWORKING MACHINERY AND WATERTIGHT BULKHEADS IN SHIPS
Samuel Bentham (1757–1831)

Bentham's woodworking machinery, which included provisions for blockmaking, was soon to be eclipsed to a large extent by that of Marc Brunel, whom Bentham generously encouraged.

Samuel Bentham combined the talents of naval architect, engineer and administrator in equal proportions and with considerable success. The son of an attorney, he was educated at Westminster School, and was subsequently apprenticed to the Master Shipwright of Woolwich Dockyard. In 1780 he went to Russia, where he eventually became a shipbuilder for the Empress Catherine. He rendered conspicuous service in the war against the Turks, and at the battle of Limars in 1788 the floating batteries of howitzers and mortars, which he had designed, destroyed eleven ships o' the line. In 1791 he returned to England and began to study the application of machine tools for woodworking. In 1793 his ideas were embodied in patent no1838, which was granted to him to cover planning with rotary cutters, a mortising machine and other

woodworking machines capable of producing timber with a curved profile suitable for the ribs of wooden ships. Two years later he was appointed Inspector-General of the Royal Dockyards, and was able to encourage greater use of machinery to perform many of the tasks previously carried out by hand. He effectively reorganised the administration of the dockyards and was instrumental in installing the first steam engine in Portsmouth Dockyard. Bentham's office was abolished in 1807, and he was then appointed a Commissioner of the Navy. To his great credit he encouraged and was largely responsible for the Admiralty's decision to adopt Marc Brunel's blockmaking machinery (ref 78) which in some respects supplanted his own earlier woodworking machines. Other developments introduced by Bentham included the use of caissons for closing the entrances of docks and the construction of watertight bulkheads to subdivide a ship's hull into a number of separate compartments. This practice was originally developed by the Chinese and was revived by Bentham during his term of

office, although it was not until iron began to supplant timber in the years immediately following Bentham's death that it was adopted to any considerable extent.

Further Reading
Gilbert, K. R., *The Portsmouth Blockmaking Machinery*, 1965

75 SLIDE REST
Joseph Bramah (1748–1814) and
Henry Maudslay (1771–1831)

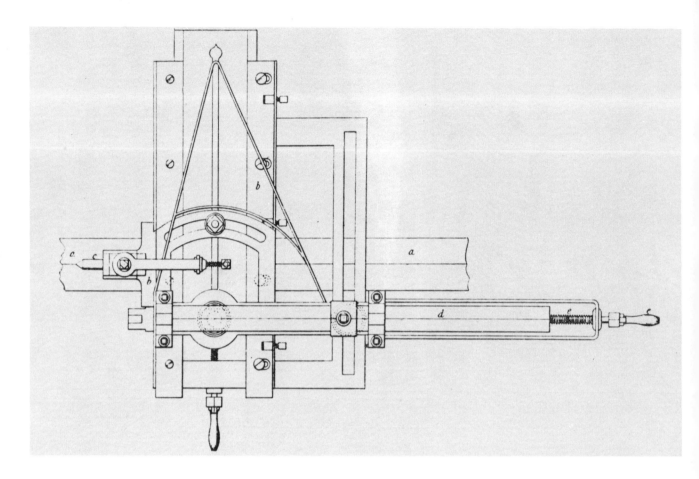

The slide rest eliminated the need to hold metal cutting tools by hand when operating a lathe. As a result tool adjustment became possible and a much higher degree of accuracy, as well as a considerably higher output, could be obtained by metal turners.

Slide rests were used in conjunction with lathes on the continent from the middle of the eighteenth century, but the first to be designed and constructed in England is thought to be that of Joseph Bramah and Henry Maudslay, which was made in 1794. Maudslay was then employed by Bramah, and it is likely that he was responsible for the manufacture and assembly of the components which were made to Bramah's design. The unit was actually a combined tailstock and tool rest, and included a longitudinal traverse and cross feed provided by manually-operated lead screws. An adjustment was also provided for taper turning. The cutting tool

was secured at the end of a square section bar which was itself mounted on the upper part of the travelling carriage. It was Bramah's intention that his slide rest should be considered as a conversion unit to be fitted in place of the original tailstock of earlier lathes where a hand rest was employed for the tool. As improved designs became available, the slide rest was incorporated as original equipment, and the demand for separate units fell away. It was nevertheless an important advance in the development of the lathe and contributed considerably to the higher standards of accuracy attained during the early part of the nineteenth century in metal working.

Further Reading

Steeds, W., *A History of Machine Tools, 1700–1910*, Oxford 1969

Rolt, L. T. C., *Tools for the Job*, 1965

76 HYDRAULIC PRESS
Joseph Bramah (1748–1814)

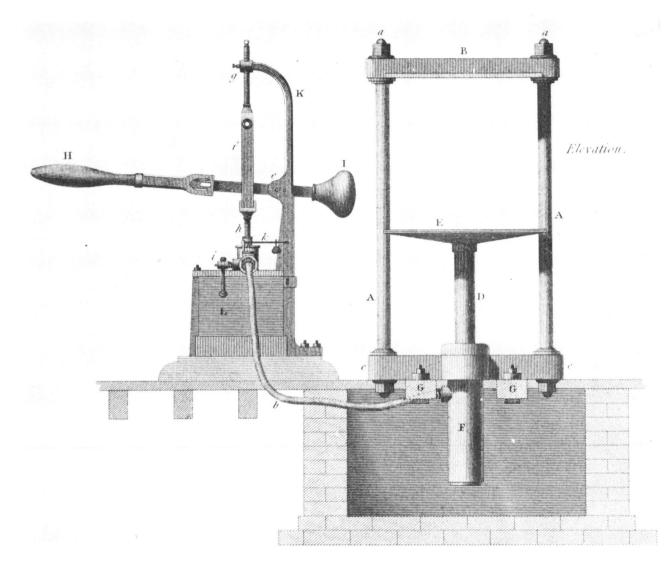

The hydraulic press provided engineers and metal workers with a tremendous new source of power, and one which was subsequently harnessed in a wide variety of industries.

The science of hydraulics was not unknown in the eighteenth century, but little progress was made in its practical application until Joseph Bramah patented his hydraulic pump in 1795. After serving his apprenticeship as a carpenter in his native Yorkshire, Bramah moved to London, where he was employed as a cabinet maker until he set up his own business. Bramah's first invention was an improved water closet; this was followed by his celebrated lock (ref 125) and by a host of other unrelated appliances, both large and small, which

demonstrated the versatility of his agile mind. Altogether he was granted eighteen patents, but none was more important than no2045, for which he applied on 30 April 1795. In addition to covering the hydraulic press, this included a device which foreshadowed modern fluid power control systems, such as the telemotor control circuit of a ship's steering gear. It was the press, however, which was first put into practical use. In Bramah's original design heavy bulks of timber were used for the frame, but these were soon replaced by wrought iron or cast steel bars as shown in the illustration. Bramah modestly described his apparatus as being 'literally no more than two pumps of different dimensions acting on each other'. This in effect was true, and Bramah was able to show that if the hand-operated pump cylinder bore was only $\frac{1}{4}$in, a load of 1 ton would produce a force of no less than 2,304 tons at a receiver having a diameter of 12in. The key component in the apparatus was the cylinder gland or seal which had to be thoroughly pressure-tight under all conditions of service. Some initial difficulties in its construction were overcome with the introduction of the self-tightening collar which has been attributed by most historians to Henry Maudslay, who was then the foreman of Bramah's Piccadilly works. A few years elapsed before the potential value of the press was fully appreciated. Until that time pressing operations in industry were conducted mainly on simple screw presses, and it was difficult for mill owners to appreciate the tremendous power of the hydraulic press. One of the first was installed by Benjamin Gott, a Leeds woollen manufacturer, in his mill in 1799, and another was sold to Joseph Ridgway of Bolton in the same year. The introduction of both these presses met with resistance from the mill workers, who feared their jobs would be jeopardised, and they lay idle for a number of years. Other early applications were for the extraction of oil from a variety of organic substances.

Further Reading

McNeil, I., *Joseph Bramah, A Century of Invention*, Newton Abbot, 1968

77 SCREW CUTTING LATHE
Henry Maudslay (1771–1831)

Maudslay's screw cutting lathe was the basic invention in the development of machine tool technology in Britain and as such exerted a momentous influence on the expansion of all engineering activities in the early years of the nineteenth century.

When Henry Maudslay left the employment of Joseph Bramah in 1797, he set up in business on his own in Wells Street, off Oxford Street, in London. Under Bramah's guidance Maudslay had already been concerned with machine tool development and it was a logical step to progress from the slide rest (ref 75) to the screw cutting lathe. It is unlikely that Maudslay was familiar with earlier lathes

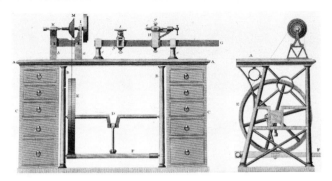

made on the continent by Vaucanson and Senot, but he was probably aware of the small screw cutting lathes used by the scientific instrument makers, of which Ramsden's had attained the highest degree of perfection. Maudslay in effect scaled up the instrument lathe but in doing this he had to overcome a number of problems, the most formidable being the production of an accurate lead screw. His machine, which is preserved in the Science Museum at South Kensington, had a bed consisting of two trianagular bars supported at each end by cast brass feet. The headstock and tailstock were carried on the rear bar with the headstock in the unusual position at the right hand end. The headstock spindle was geared to the lead screw which was

located between the bars of the bed and supported at its left hand end by a bracket carried by the front bar. The saddle with the tool holder was supported by both bars and the tool adjustment incorporated several of the features initially developed for the Bramah/Maudslay slide rest of 1794. Proof that Maudslay was one of the greatest craftsmen of his age is given by the construction of his first lead screw which had to be cut by hand. The material was either hard wood or soft metal and the thread was cut initially by an inclined knife and no doubt finished by filing. This first lead screw was then fitted to the lathe and used to machine an identical counterpart in iron or steel which replaced the original. Although Maudslay is chiefly remembered as an engine maker, he continued to work on the development of the lathe and by 1810 had introduced a number of refinements which he incorporated into a new model that fetched a selling price of £200.

Further Reading

Steed, W., *A History of Machine Tools, 1700–1910*, Oxford, 1969
Rolt, L. T. C., *Tools for the Job*, 1965

78 BLOCKMAKING MACHINERY
Marc Brunel (1769–1849)

The Portsmouth blockmaking machinery is of immense historical significance since it was the first example of large scale mass production of a component using a series of specially-designed machine tools. Brilliant in conception and execution, it was an outstanding success and saved the Admiralty enormous sums of money until sail eventually gave way to steam.

Although the forty-five machines which are collectively known as the Portsmouth blockmaking machinery were not built until the early years of the nineteenth century, there is no doubt that they were conceived and certain preliminary design work was undertaken in the closing years of the eighteenth century. The plan for manufacturing ships' blocks in vast quantities by mass production techniques was put forward by Marc Brunel, when occupying the post of Chief Engineer of New York. Brunel's career in America was brief but meteoric. He arrived in the United States as an emigré royalist in 1794, after service in the French navy, and sailed back

across the Atlantic to England early in 1799. He took out a British patent, no2478, for his blockmaking machinery in 1801 and through the good offices of his brother-in-law, who was Under Secretary to the Navy Board, he obtained the support of Sir Samuel Bentham (ref 74) who then occupied the post of Inspector-General of Naval Works. Bentham persuaded the Admiralty to accept Brunel's proposals in April 1802, and Henry Maudslay, who had previously collaborated with Brunel in the construction of models, was entrusted with the task of making and erecting the full size machines in Portsmouth Dockyard. The machines were made in three stages between 1803 and 1807. Altogether there were twenty-two different types of machine; they were erected on two floors of a building and were driven by shafting linked to one of the two steam engines previously introduced into the dockyard by Bentham. The elm logs, which provided the basic material for the block shells, were successively sawn, bored and mortised, and

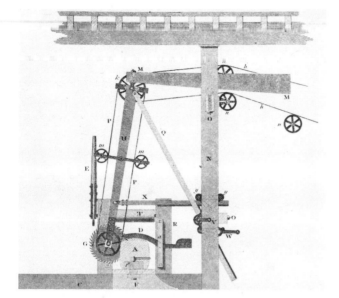

the shells were finally given the traditional oval shape in a shaping engine. The sheaves or pulleys were made from lignum vitae and were similarly formed on a separate series of special machines. Coaks made of bell metal were inserted in the sheave and an iron pin was used to locate the sheave in position in the shell. The output of the blockmaking machinery was prodigious. In 1808 130,000 blocks of varying sizes were produced, and two years later all contracts with private suppliers were cancelled. The capital cost of the machinery is said to have been recovered in three years from the savings in costs of production—a claim which is not difficult to credit since ten unskilled men working the machines could replace a hundred and ten skilled craftsmen.

Further Reading

Gilbert, K. R., *The Portsmouth Blockmaking Machinery*, 1965

CIVIL ENGINEERING

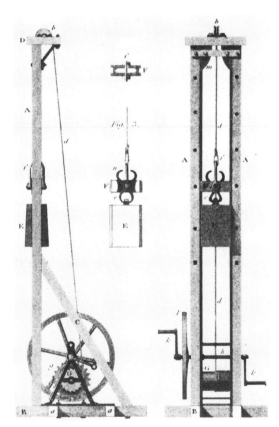

79 PILE DRIVING MACHINE
Charles Valoué

Pile drivers, such as the type invented by Valoué, came into service during the eighteenth century for use in bridge building and for harbour works—two aspects of civil engineering in which appreciable advances were made.

The most effective pile driver constructed during the eighteenth century is generally considered to have been that invented by Valoué, a London clockmaker, during the 1730s. Valoué's machine appeared at a most opportune moment, and a version of it was used in the construction of Westminster Bridge during the years 1738–1750. The pile driver in question was built by Valoué to the order of Charles Labelye, the engineer in charge of the bridge construction. It was completed in September 1738 and, unlike the version shown in the illustration, it was driven by two or three horses operating a gin. A system of gear-wheels, ropes and pulleys was employed to raise the ram which weighed 1,700lb. The most ingenious part of the mechanism was the spring-loaded gripping device which could move up and down to grip the ram and opened automatically when the ram was at the top of its stroke. By modern standards the speed of operation was slow; with two horses it was

capable of 48 strokes per hour and 70 strokes per hour with three. Labelye ordered the pile driver to be mounted on a barge, and it was used initially for sinking a series of piles around the bridge works to keep the river traffic away from the scene of operations. The piles were made from trunks of fir trees and were about 13–14in square and 34ft long. Their points were sheathed with iron and iron hoops were also used to reinforce the top ends. Valoué's machine was described by J. T. Desaguliers in his publication *Cours de Physique Experimentale* as about five times as effective as any pile driver that he had previously seen. Other civil engineers were obviously in agreement, and a Valoué machine was soon exported across the Channel, where it was used in the construction of the Seve Bridge, on the road from Paris to Versailles. This particular machine is known to have been worked by men, and manual power seems to have been preferred to horse power for many of the later models.

Further Reading
Jensen, M., *Civil Engineering around 1700*, Copenhagen, 1969

80 ROAD MAKING TECHNIQUES
Pierre Tresaguet (1716–80?)

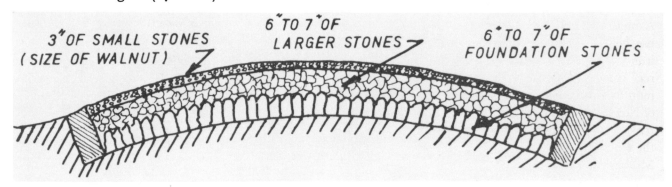

The efficient system of roads created by Tresaguet and his assistants in France gave the French armies in the Napoleonic wars a degree of mobility on their own territory that was much greater than that enjoyed by their enemies.

Pierre Tresaguet constructed the first well-engineered road of modern times, and his work preceded that of McAdam in Britain by about

forty years. France, as the leading military power in Europe, required an efficient road system as a matter of policy, and by 1720 had already created a government department responsible for the building and maintenance of roads and bridges. Towards the middle of the century the Ecole des Ponts et Chaussées was founded, which trained a nucleus of civil engineers who greatly improved the existing French roads without introducing any fundamental change in road construction. It was left to Tresaguet in the last creative decades of the French monarchy to develop the techniques which, with modifications, have endured to the present day. Drainage and durability were the main principles behind Tresaguet's ideas. He used a foundation layer of flat stones set on edge about 6in high. This was covered by a further layer of smaller stones, with a final 3in layer of stones which were about the size of a walnut. When these were bedded in they formed a hard waterproof surface, and the slight camber caused the rainwater to run off into the side ditches.

Further Reading
Pannell, J. P. M., *An Illustrated History of Civil Engineering*, 1964

81 LIGHTHOUSE
John Smeaton (1724–92)

John Smeaton's lighthouse on the Eddystone rock, 11 miles south of Plymouth is generally considered to be the first modern example of a type that was afterwards adopted by other maritime nations.

Although lighthouses had been built since Classical times, their usefulness was often limited because of the inadequacy of their construction. Smeaton, who was only thirty-one when he was recommended in 1755 as the designer of the third Eddystone lighthouse, wisely rejected most of the ideas adopted by his predecessors. He decided to use stone in preference to timber and based the design of his tower on the trunk of an oak tree, thus spreading the foundations over the greatest possible area and at the same time offering the least resistance to the sea and wind. Smeaton approached the task with his customary thoroughness and dedication. He visited the reef and made a careful examination of the rock strata. A model was then made in which his proposals to use blocks of stone with interlocking dovetails and iron cramps were clearly shown. Work began at the reef early in August 1756 and for the next three months, until adverse weather set in at the beginning of November, the dovetail recesses in the foundation rocks were cut out. During the winter of 1756–7 the stones for the tower were prepared in a yard at Millbay in Plymouth. Portland stone was selected for the bulk of the masonry, with granite for the outer cladding. Work began again at the reef in June 1757 and by the end of the summer nine courses were in place. The same programme was repeated over the next twelve months, with the stones being carefully prepared ashore during the winter and erected during the fair weather of the summer. In 1758, in spite of many interruptions, the tower was raised to the twenty-ninth course and

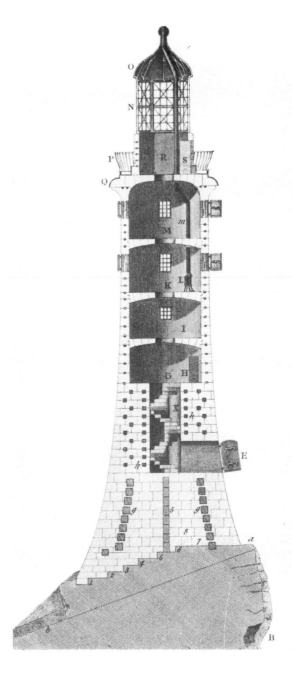

finally completed in the following year. The main stone column was 70ft high with a diameter of 28ft at the base and the lantern added a further 28ft to the structure. The light, which was first brought into service on 16 October 1759, consisted of twenty-four candles in a chandelier. Smeaton rejected the use of oil lamps as they deposited too much black smoke on the glass of the lantern. The soundness of Smeaton's ideas was proved by the endurance of his lighthouse which remained in service until about 1880, the structure only then being replaced because of undermining of the reef on which it stood. The upper part was dismantled and re-erected on Plymouth Hoe and the original base was left on the reef as a permanent memorial.

Further Reading
Pannell, J. P. M., *An Illustrated History of Civil Engineering*, 1964

82 ROAD BUILDING
John McAdam (1756–1836)

McAdam—the first specialist in highway engineering—introduced road making techniques which gave Britain a system of roads that was more than adequate to meet the demands of the horse drawn traffic throughout the nineteenth century.

In his early life John McAdam gave little indication that he might become the most renowned of all British highway engineers and thereby add a new word to the English language. He was born at Ayr in Scotland, but at fourteen went to live in New York in the care of an uncle who was in business as a merchant. McAdam remained in New York until after the War of Independence; while in America he was able to accumulate a considerable fortune through his position as British agent for the sale of prizes. He returned to Scotland and purchased an estate at Sanhrie in Ayrshire. His affluence permitted him to pursue a number of interests and it was during this period in Ayrshire between 1785 and 1798 that he began to experiment in road construction—an activity which no doubt followed from his appointment as Deputy Lieutenant and Road Trustee of his native county. McAdam's ideas on road making were based on one fundamental principle—efficient drainage. This was a view which he shared with his celebrated predecessor, Tresaguet in France (ref 80) and with his contemporary, Thomas Telford. Unlike Telford, however, who favoured a heavy stone foundation, McAdam considered that the weight of the traffic should be borne by the sub-soil which he believed would support any load provided it was preserved in a dry state. To achieve this he raised the road above the water table and produced a hard impervious surface with a camber to assist drainage. His

MACADAM'S SPECIFICATION

2" LAYER OF 1" BROKEN STONES.

4" LAYER OF 2"-2½" BROKEN STONES. (6oz STONES)

4" LAYER OF 2"-2½" BROKEN STONES. (6oz STONES)

SUBSOIL.

foundation stones were gauged in a 2in ring and their individual weight was limited to 6oz. The stones were laid in three layers, each being about 4in deep. One layer of stones was packed down and rammed to the correct camber before the next was laid. He relied on consolidation of the relatively small stones by the iron tyres of coaches and wagons to give the required surface. McAdam's methods were less expensive than those of Telford on first cost, and were widely adopted in Britain. He continued with his experiments after leaving Scotland and was able to put many of his ideas into practice on his own account when he became, successively, Surveyor-General of Bristol roads in 1815 and General Surveyor of Roads to the Government in 1827.

Further Reading

Pannel, J. P. M., *An Illustrated History of Civil Engineering*, 1964

MARITIME

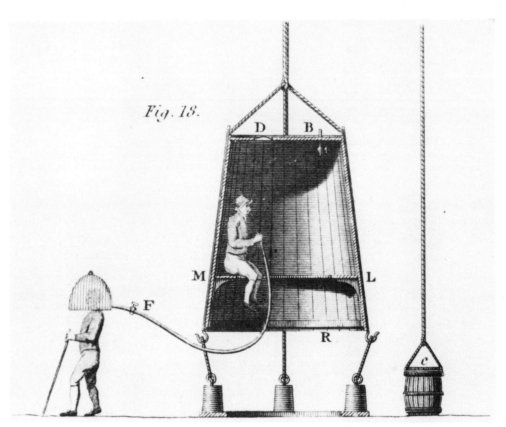

Fig. 18.

83 DIVING BELL
Edmund Halley (1656–1742)

Tentative steps were taken during the eighteenth century to explore the world beneath the sea, and a number of primitive diving bells were developed which were the true antecedents of the modern underwater laboratories.

The celebrated astronomer Edmund Halley was also concerned with exploring the world under water and was one of several inventors of diving bells. He gave a detailed description of his apparatus in a paper presented to the Royal Society in September 1716, when he was himself Secretary of the Society. In earlier versions of the diving bell no provision was made to replenish the air, with the result that it became foul and the period that could be spent underwater was limited. Halley was aware of this defect, although another half century was to pass before it could be explained by the discoveries of Priestley and Lavoisier. He merely described the air as 'becoming heated and unfit for respiration'. His solution was to send down to the sea bed close to the diving bell weighted barrels which were empty and watertight. A seat was provided for the diver within the diving bell which was also located in position by weights, and a hose with a discharge cock at the end was led from the barrels to the interior of the bell. By turning on the cock the diver could mix the fresh air from the barrels with the air already present inside the bell. Another refinement suggested by Halley was the provision of individual head pieces which could be used by divers working outside the bell, the head piece being linked by a flexible hose to the air space in the bell. Halley put his theories into practice, and on one occasion with four companions spent an hour and a half submerged to a depth of ten fathoms. There is little evidence to indicate that his diving bell was put to much practical use, but one contemporary print does show a bell similar to Halley's being employed for the task of raising cannon from a sunken ship. It was left to John Smeaton, however, to use a diving bell for the first time on harbour construction work, at Ramsgate in 1774.

Further Reading
De Latil, P., and Rivoire, J., *Man and the Underwater World*, 1956

84 STEAM TUG
Jonathan Hulls (1699–1755)

Jonathan Hulls' steam tug was a failure and no proof has ever been presented that it moved even a few yards under power. However the arrangement of the paddles at the stern was simpler and more practical than most other methods of propulsion proposed by other steamboat pioneers later in the eighteenth century.

One of the problems which became increasingly acute as England's royal and merchant navies expanded was that of moving sailing ships in and out of harbour when becalmed. Jonathan Hulls, a clock repairer of Chipping Camden in Gloucestershire, conceived the idea of fitting a Newcomen steam engine in a small boat and arranging it to drive a pair of paddle wheels mounted at the stern. The boat or tug would then tow sailing ships with their sails furled to the required position. In 1736 Hulls obtained the support of a Mr Freeman of Batsford Park near Chipping Camden who gave him about £160 to develop his invention and in the same year in December he obtained a patent—no556. Translating his proposals into practice was however beyond his capabilities which is hardly surprising since the atmospheric steam engine at that time was designed for vertical rather than rotary motion. Hulls is known to have made a series of drawings which he sent to the Eagle Foundry at Birmingham where the large castings and forgings were made. He selected the River Avon at Evesham for the site of his experiment and a boat was built, probably in an Evesham boatyard. The only reference to the trial, which was carried out some time in 1737, is cryptically that it was a failure. Hulls' financial support was withdrawn and he abandoned the project after making an unsuccessful attempt to obtain further patronage through the publication of a pamphlet in London under the somewhat extended title of *Description and Draught of the new-invented Machine for carrying vessels or ships out of or into any Harbour, Port or River against wind and*

G

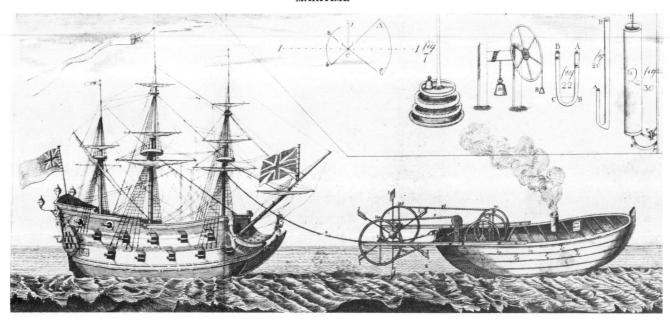

tide or in a calm. It is perhaps ironic to record that when steam navigation was first introduced into the Royal Navy nearly a hundred years after Hulls' proposals had proved a failure, one of the main duties of a steam warship was to tow the line o' battle ships in and out of harbour during adverse sailing conditions.

Further Reading

Spratt, H. P., *Marine Engineering*, Science Museum Catalogue, 1953

Rowland, K. T., *Steam at Sea*, Newton Abbot, 1970

85 STEAMBOAT
Claude de Jouffroy (1751–1832)

Jouffroy designed and built the first steamboat to operate under its own power—albeit only briefly. This was a major achievement in the history of transport but France was denied a lead in steamboat development by petty bureaucracy and the outbreak of the French Revolution.

After Hulls' abortive attempt in England, more than a quarter of a century elapsed before two French aristocrats, Le Comte Joseph d'Auxiron and Le Chevalier Charles Monnin d'Eollenai, tried to build a similar steamboat in 1772 on the banks of the Seine. They were no more successful than the English clockmaker, their vessel sinking mysteriously before a trial could be held. A third nobleman, the Marquis Claude de Jouffroy d'Abbans, was also associated with the project and after an interval of several years Jouffroy began experiments on his own accord at Baume-les-Dames on the River Doubs near Besançon. Jouffroy had previously studied the paddling movements of certain aquatic birds, and he designed an apparatus comprising a series of rods, about 8ft long. It was his intention to actuate the rods, which carried paddles at the free ends, by means of a steam engine. Trials were held on the River Doubs in June and July 1778 but the paddling mechanism proved defective. A second boat was built at Lyons and in the latter enterprise Jouffroy was joined by Charles d'Eollenai. The vessel, which was called the *Pyroscaphe*, was fitted with a horizontal double-acting engine with a single cylinder measuring 25.6in diameter and having a stroke of 77in. The engine was enclosed within the boiler brickwork and Jouffroy dispensed with his original paddle system in favour of the more conventional paddle wheels mounted on either side of the craft. On 15 July 1783 the *Pyroscaphe* steamed upstream for 15 minutes on the River Saône—the first time in history that a vessel had been moved by the power of steam. The event was witnessed by a Commission de Savants, and Jouffroy confidently expected to receive sufficient financial backing to continue with his experiments and eventually form a company. He applied to the French government for aid and sent with his application a scale model of

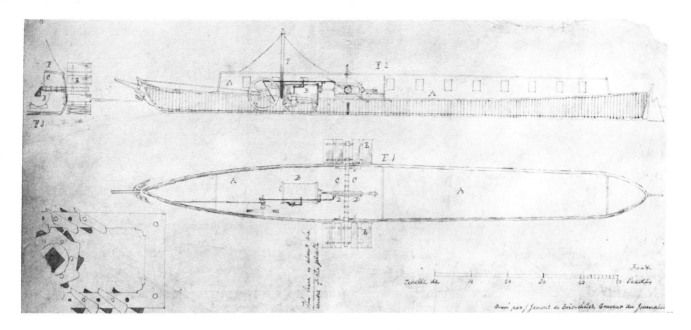

the *Pyroscaphe* to support his claim. The matter was referred to the Academy of Science who in turn appointed a Commission. Further delays followed while the Commission sought the advice of Jacques Perier who had himself failed to build a steamboat on the Seine a few years earlier. Whether motivated by jealousy or lack of evidence, Perier influenced the Commission to declare that the trials at Lyons had not been decisive. Jouffroy was refused permission to form a company and abandoned all work on the project when financial support was not forthcoming. When the French Revolution broke

out, he went into premature retirement. He witnessed Robert Fulton's early steamboat experiments on the Seine and at the age of sixty-five he emerged to head a company which built a steamboat at Petit-Bercy near Paris. This venture was also a failure and Jouffroy died almost forgotten in the Hôpital des Invalides in 1832.

Further Reading

Spratt, H. P., *Marine Engineering*, Science Museum Catalogue, 1953

Rowland, K. T., *Steam at Sea*, Newton Abbot, 1970

86 STEAMBOAT
John Fitch (1743–98)

The world's first commercial steamboat service was operated during the summer of 1790 on the Delaware River by John Fitch and his partner but it could not compete effectively against the stage coach.

John Fitch, a rival and contemporary of Rumsey, was one of the small band of eighteenth-century pioneers whose work formed the foundation of commercial steamboat development by men such as Robert Fulton and Henry Bell in the early years of the nineteenth century. Fitch and the other Americans, who were seeking primarily to introduce a means of steam-driven transport on the rivers of the eastern seaboard of the United States, were handicapped by the lower standard of craftsmanship

that existed there compared with that of western Europe. It is proof of Fitch's tenacity that he eventually triumphed over these handicaps although he owed much to the assistance of his partner, Johann Voigt, a former German watchmaker. Their first experimental craft was tried on the Delaware River on 27 July 1786. It was powered by an engine with a 12in diameter cylinder equipped with a separate condenser. The most interesting feature was the method of propulsion which consisted of two banks of paddles, six in each bank on either side of the boat. The paddles were actuated by cranks driven from the engine and the apparatus was a crude but nevertheless logical attempt to translate the action into mechanical terms of a

group of Indians paddling a canoe. It was only partly successful, a speed of three miles per hour being attained for a short distance. Further trials were carried out in 1787 but it was not until 12 October 1788, when a twenty-mile trip was made from Philadelphia to Burlington, that any real progress was achieved. On this occasion thirty passengers were carried at an average speed of 5.5 miles per hour. During the next eighteen months Fitch and Voigt experimented with a new engine cylinder and the paddles were relocated at the stern. In the spring of 1790 they appeared to be on the brink of success when a new condenser, similar to Watt's jet condenser, was fitted. This increased the engine efficiency and, on trials, the speed was increased to seven miles per hour. During the summer the partners operated a steamboat service between Philadelphia and Trenton—the first com-

mercial service of its type in the world—but although breakdowns were rare the service did not pay. Fitch had overlooked the competition of the stage-coach which was faster, although more expensive. In 1793 Fitch travelled to Europe where a French patent was obtained on his behalf but in the chaotic conditions then prevailing further development work was out of the question. He returned to Boston in 1794 and is reported to have carried out experiments in screw propulsion on the Collect Pond, now part of the City of New York, two years later. He died in Kentucky in 1798.

Further Reading

Flexner, J. T., *Steamboats Come True*, New York, 1944
Spratt, H. Philip, *The Birth of the Steamboat*, 1958
Rowland, K. T., *Steam at Sea*, Newton Abbot, 1970

87 STEAMBOAT
William Symington (1763-1831)

William Symington and his patron, Patrick Miller, were the true pioneers in Britain of the steamboat, although the first commercial service did not operate until the beginning of the nineteenth century.

James Watt, as an advocate of low pressure steam, showed no inclination to adapt his successful steam engine for maritime transport but other innovators and engineers did not share his reserva-

tions. Indeed two of Watt's fellow countrymen, Patrick Miller and William Symington, built a small but successful steamboat in 1788 while the Soho foundry was allegedly enjoying a monopoly in the production of condensing engines. Patrick Miller, who provided the finance, was a retired Edinburgh merchant with an interest in naval architecture. He originally experimented with twin-hull vessels driven by paddles—the driving force being the muscle power of the crew who operated the paddles through a capstan. Miller decided to substitute a steam engine in place of the capstan and engaged a young civil engineer, William Symington, to build an engine for him. Symington's engine worked on the Newcomen or atmospheric principle but it included a separate condenser and was really an infringement of Watt's patent. It consisted of two 4in cylinders which were cast in brass; the pistons were connected by chains to a drum which was turned alternately in opposite directions. Drive to the paddle wheels was transmitted from the drum through a complex mechanism of pulleys and pawls, the paddles being located between the twin hulls as in Miller's earlier vessels. The engine and paddle-drive appear to have worked tolerably well when trials were held on a lake at Miller's estate at Dalswinton in Dumfries. A speed of about five miles per hour was attained and

Miller was encouraged to commission Symington to build a larger engine so that they could continue their experiments. The cylinders for the second vessel measured 18in diameter but the same system of driving the paddles was retained. Difficulties were encountered with the driving chains and it was at this juncture that Watt was consulted for his advice. This was probably the first time that the inventor had been acquainted with the details of Symington's engine and his reply to Miller on 24 April 1790 was anything but encouraging. He wrote of Symington's attempt 'to evade our exclusive privilege,' and since the Soho partners had shown no reluctance to engage in litigation to protect their patents both Miller and Symington abandoned further experiments with steam. After Watt's patent had expired in 1800, Symington built the engine for the *Charlotte Dundas*, a steam tug which operated successfully for a time on the Forth-Clyde canal, but a combination of circumstances including the death of the Duke of Bridgewater, who had intended to order eight additional steam tugs, prevented him from enjoying the success he deserved.

Further Reading
Spratt, H. P., *The Birth of the Steamboat*, 1958
Rowland, K. T., *Steam at Sea*, Newton Abbot, 1970

88 STEAMBOAT
James Rumsey (1743-91)

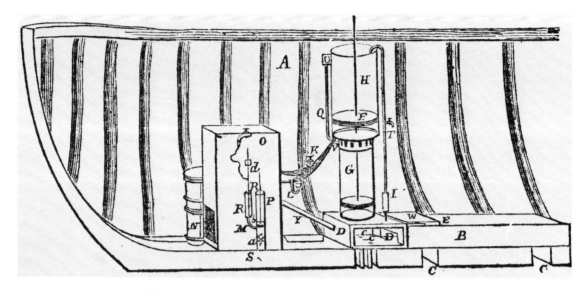

James Rumsey, an early American steamboat pioneer, introduced water jet or hydraulic propulsion. He also patented a pipe boiler which was the first special marine boiler.

James Rumsey was a contemporary of John Fitch (ref 86) and the two men were rivals for the honour of building the first successful American steamboat. Rumsey first began experimenting with steam in 1785 on the River Potomac and immediately devised the system of hydraulic propulsion which he used in all his craft. The propelling apparatus consisted of two cylinders placed one above the other, with two pistons connected by a common rod. The lower cylinder acted as a pump and was in turn connected to a series of valves fitted in the bottom of the boat. Steam from the boiler was admitted beneath the piston in the upper cylinder and forced it to move up. The lower piston also moved up and water was drawn in through the valves at the bottom. A separate condenser was fitted and the steam was exhausted to it when the upper piston reached the top of its stroke. Atmospheric pressure then forced the piston down, and water in the lower cylinder was pumped out through a trunk at the stern of the boat, the reaction of this pulsating stream driving the boat forward. Rumsey's first boiler was a simple iron pot, but he considered that this was too heavy and substituted a pipe boiler which was probably his major contribution to steamboat development. The pipe boiler was

a model of compactness, standing only 3½ft high, and yet it had a heating surface of 62sq ft. Inside the boiler was a single coil of 2in iron pipe, through which the water circulated. Work on the vessel was dishearteningly slow; abortive trials were held in the spring of 1786 and it was not until December 1787 that some success was achieved, when the craft moved against the current at a speed of three miles an hour carrying a burden of two tons in addition to the weight of the machinery. In the following year Rumsey journeyed to England to secure a British patent for his apparatus. He was financed on this venture by the Rumseyan Society of whom Benjamin Franklin was a prominent member. Rumsey achieved his immediate object, being granted patent no1673 in 1788, but in other respects practical success eluded him. A steamboat to his design named the *Columbian Maid* was built at Dover and fitted with a pump with a cylinder of 24in diameter equipped with a piston that made 20–22 strokes per minute. Trials were held in December 1792, but before they could be completed, Rumsey died of a stroke while giving a lecture at the Society of Arts in London.

Further Reading
Flexner, J. T., *Steamboats Come True*, New York, 1944
Rowland, K. T., *Steam at Sea*, Newton Abbot, 1970
Spratt, H. P., *Marine Engineering* Part II, Science Museum Catalogue, 1953

89 SCREW PROPELLER
William Lyttleton

Lyttleton's screw propeller was the result of another attempt to solve the problem of moving sailing vessels in and out of harbour or when becalmed.

The increase in trade in the second half of the eighteenth century resulted in a number of problems stemming mainly from the lack of shipping space and these were frequently accentuated by vessels being becalmed, often within sight of land. William Lyttleton, a London merchant, who may well have suffered a financial loss himself under such circum-

stances, took out a patent, no2000, in 1794 to overcome these vexing delays. His device was described as an 'aquatic propeller' and consisted of a triple threaded screw supported in a frame. It was Lyttleton's intention that it should be suspended beneath the stern aft of the rudder or at the bow or the sides. In the specification it was shown to be manually driven from a winch on deck, which was connected by ropes and pulleys to the screw shaft. It was suggested, however, that a small steam engine would be more suitable for this purpose. There is no evidence that the device was put to much practical use and its effectiveness under manual operation would be doubtful. A similar device patented by Edward Shorter in 1800 was fitted experimentally to the transport vessel *Doncaster* and tried out in Gibraltar Bay, a speed of 1.5km being attained when the vessel was becalmed with eight men operating the driving capstan. The multiple start screws, such as those designed by Lyttleton and Shorter, did however provide a basis for the work of John Ericsson and Francis Smith thirty years afterwards.

Further Reading

Spratt, H. P., *Marine Engineering*, Science Museum Catalogue, 1953
Rowland, K. T., *Steam at Sea*, Newton Abbot, 1970

90 DIVING SUIT
Klingert

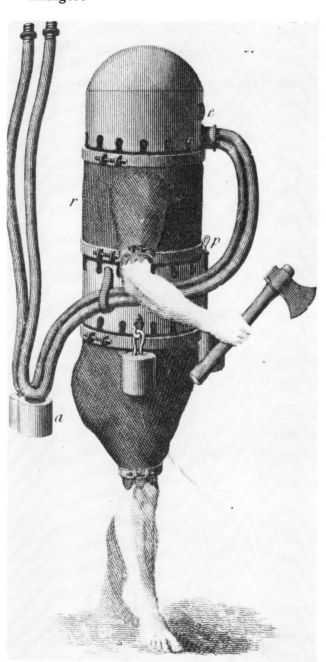

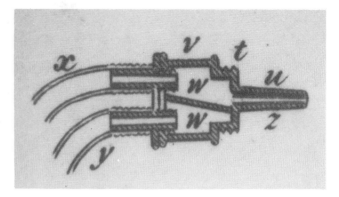

Klingert's diving suit was not sufficiently flexible to give the wearer much freedom of movement, but it incorporated features which were subsequently adopted for more successful apparatus in the nineteenth century.

The diving suit devised by the German Klingert and tested before a large crowd in the River Oder in 1797 showed several significant advances over the apparatus of Halley and Spalding. Klingert was handicapped, however, by not having a suitable material for the non-metallic parts of the suit. He used leather for the breeches and sleeves, the ends being tied tightly round the legs and arms of the wearer to prevent water from getting inside. This restricted circulation and movement, and the leather itself did not permit much freedom of action. Divers in fact were not able to achieve ease of movement until sheet rubber became available during the nineteenth century. Unlike many earlier models Klingert's suit was served with both an air supply and a discharge line, and a simple non-return valve was fitted to the mouthpiece.

Further Reading

De Latil, P. and Ricoire, J., *Man and the Underwater World*, 1956

91 SUBMARINE
Robert Fulton (1765–1815)

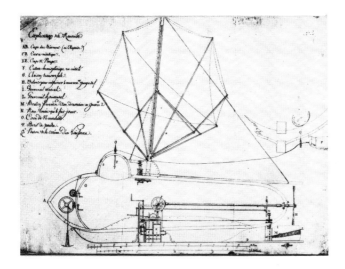

Nearly a hundred years were to elapse after the brief career of the Nautilus before the submarine became a practical reality but Fulton's primitive craft possessed many fundamental features that were to be incorporated in the early submersibles of the late nineteenth century.

Robert Fulton eventually achieved immortality with his steamboat *Clermont* on the Hudson River in 1807, but he also had his share of disappointments and failures during his eventful career, particularly in the twenty years that he spent in Britain and France between 1787 and 1807. Fulton was not the first eighteenth-century inventor to turn his attention to an undersea craft. Indeed his compatriot, David Bushnell, had achieved some degree of success with a one-man submarine with which he attempted to attack British warships blockading American ports during the Revolutionary War. Fulton's submarine was also conceived to aid a Revolutionary cause— that of the French Directorate who were engaged in an unequal maritime struggle against Britain. From 1797 until almost the end of 1799 Fulton, who had left England because of his republican sympathies, sought to convince the French authorities of the feasibility of his plan to attack the blockading British fleet with a submarine—one that would be far more effective than the primitive craft built by Bushnell. It was not until Napoleon had been appointed First Consul that the French authorities began to consider Fulton's ideas favourably and he was encouraged to translate his plans into reality. Work began in the closing weeks of the century and by April 1800 the copper hull was completed. By July the craft was finished and Fulton with two companions carried out two trial dives at Le Havre, remaining submerged eight and seventeen minutes respectively. The vessel contained a number of remarkably sophisticated features, many of which were eventually incorporated in the various proto-types built at the end of the nineteenth century. Indeed it was probably only the absence of a suitable prime mover that prevented Fulton from attaining complete success. The *Nautilus*, as the craft was named, was in fact driven by a propeller, which was hand-operated through gearing when submerged, and by a sail attached to a retractable mast when on the surface. The hull was 21ft 4in long and 7ft in diameter, and was fitted with a heavy keel that could be detached in an emergency. Buoyancy was controlled by pumps and ballast tanks, as in a modern submarine, and a horizontal rudder was fitted to assist movement in a vertical plane. A hemispherical conning tower was provided for'ard with glass look-out scuttles. The career of the *Nautilus* was, from a British point of view, commendably brief and the only two British ships encountered by Fulton moved out of range before he could attack them. In the winter of 1800–1 the *Nautilus* was modified and refitted at Brest but a change of policy, following the appointment of a new Minister of Marine, led to the withdrawal of French financial support and Fulton abandoned the project in favour of other activities.

Further Reading
McNeil, I., 'Robert Fulton—Man of Vision', *Engineering Heritage*, Vol 2, 1966

CHEMISTRY

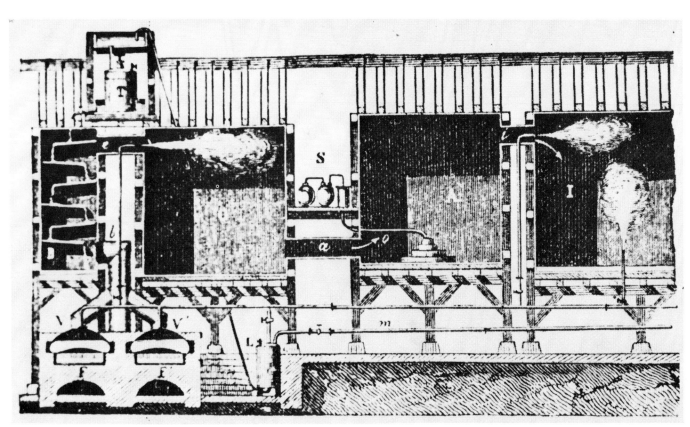

92 SULPHURIC ACID MANUFACTURE
John Roebuck (1718–94)

The invention of a process for the manufacture of sulphuric acid cheaply and in bulk had far-reaching repercussions in a variety of industries, notably textile bleaching, soap manufacture and metal refining.

John Roebuck was originally a medical practitioner who carried out chemical experiments in his spare time. He was born in Sheffield, the second son of a cutler of some means, and was educated at Edinburgh University and a medical school in Leyden where he received his MD degree. In 1744 he set up in practice in Birmingham where he met Samuel Garbett, an entrepreneur who had begun life as a brass worker. Roebuck and Garbett started a small metal refinery in Steelhouse Lane, Birmingham, to supply gold and silver to the jewellery trade, and at the same time they acted as consultant chemists to other manufacturers in the city. One of their activities involved the production of sulphuric acid which was then made in glass chambers of limited size. Roebuck could see that the cost of the end product was artificially high due to the frequent breakages of the chambers and their relatively small capacities. He proposed the use of lead-lined wooden chambers and a pilot plant was constructed which consisted of thirty chambers each measuring 8ft by 6ft by 4ft. The lead chamber process, as it came to be known, was an immediate success and the price of sulphuric acid fell dramatically. The partners could see that Birmingham was an unsatisfactory site for large scale production and, accordingly, moved to Prestonpans near the Firth of Forth three years later, where they built a plant with 108 chambers. There was a sustained demand for sulphuric acid in Scotland for use in the bleaching of linen, and their business prospered. Both men were involved in the founding of the Carron Company in 1759, the first major ironworks in Scotland, and Roebuck towards the end of his life became the patron of James Watt—an association, however, that brought little reward to either man. The lead chamber process remained in use for well over a century and the size of the chambers grew from a few hundred cubic feet to the capacity of a small concert hall. The availability of cheap sulphuric acid in bulk, together with the cheap iron made by Cort's puddling process, were the principal factors behind the rapid expansion of British industry from about 1790 onwards.

Further Reading

Hardie, D. W. F. and Pratt, J. D., *A History of the Modern British Chemical Industry*, Oxford, 1969
Chapman, S. D. and Chambers, J. D., *The Beginnings of Industrial Britain*, 1970

93 DISTILLATION APPARATUS
Peter Woulfe (1727–1805)

Woulfe's apparatus made a notable contribution to the development of experimental chemistry and was used by many chemists of the day, including Priestley and Lavoisier.

Peter Woulfe made many notable contributions to the development of chemistry in the eighteenth century, but he is chiefly remembered for the glass distillation apparatus which he devised and which still bears his name. Woulfe was elected a Fellow of the Royal Society in 1767, and in November of that year he presented his classic paper on *Experiments on the Distillation of Acid and Volatile Alkalies*, in which he described his apparatus for passing gases through liquids for the preparation of ammonia, hydrochloric acid and nitric acid, and for the distillation and collection of acids in fractions. The importance of Woulfe's apparatus was immediately recognised, and he was awarded the society's Copley Medal in 1768. Woulfe divided his time between London and Paris, and many leading French chemists began to use Woulfe's bottle and other parts of his apparatus from about 1773 onwards. Lavoisier recognised his debt to Woulfe in his famous treatise of 1789, and leading chemists on both sides of the Channel adopted his equipment for many aspects of their work, until it became standard equipment in all laboratories.

Further Reading

Partington, J. R., *A History of Chemistry*, Vol III, 1962

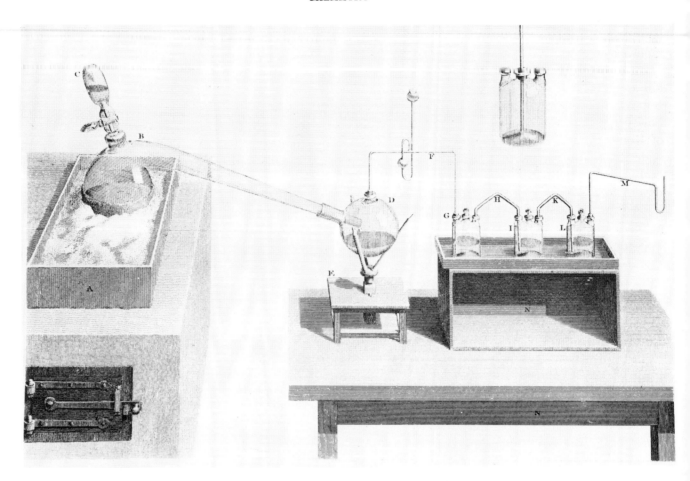

94 BLEACHING AND DYEING TEXTILES
Claude Berthollet (1748–1822)

The introduction of chlorine as a bleaching agent for cloth simplified the process and reduced the time taken for bleaching by a significant amount.

Only a relatively few years elapsed between the isolation of the gas chlorine by the Swedish chemist Scheele in 1774 and its adoption in a major industrial process—the bleaching of textiles. Until the latter part of the eighteenth century it was the usual practice to bleach cloth by immersing it in sour milk and then leaving it to dry in sunlight, the process being repeated over a period of several months. The French chemist Berthollet demonstrated the bleaching properties of chlorine in 1785; he discovered that the gas could be dissolved in potassium hydroxide to form a solution which he called *eau de Javelle*. This overcame the problem of bringing the chlorine into contact with the cloth; but the odour was persistent and unpleasant, and it remained for Berthollet's successors to devise

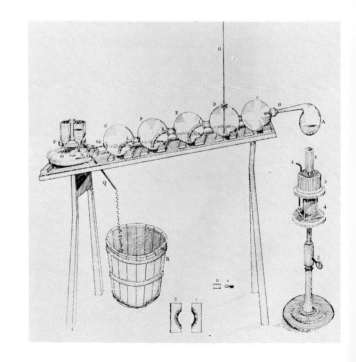

methods of removing it. Two of Berthollet's assistants attempted to persuade the French textile industry to adopt his methods; they met with varying success—at Lille there was little opposition but at Valenciennes the bleachers in the town were bitterly opposed to the new techniques. In 1791, however, a machine was invented by Descroizelles of Rouen which could bleach textiles on a large scale. It used *eau de Javelle* and Descroizelles generously called his machine the *Berthollimetre*. Further developments took place in Britain when Charles

Tennant discovered a bleaching powder made by uniting chlorine with slaked lime which he proceeded to manufacture in large quantities. Berthollet also made a number of improvements in the dyeing of textiles. He published an important treatise on the subject in 1791 which was translated into English in the same year.

Further Reading

McCloy, S. T., *French Inventions of the Eighteenth Century*, Kentucky, 1952

95 BLEACHING POWDER
Charles Tennant (1768–1838)

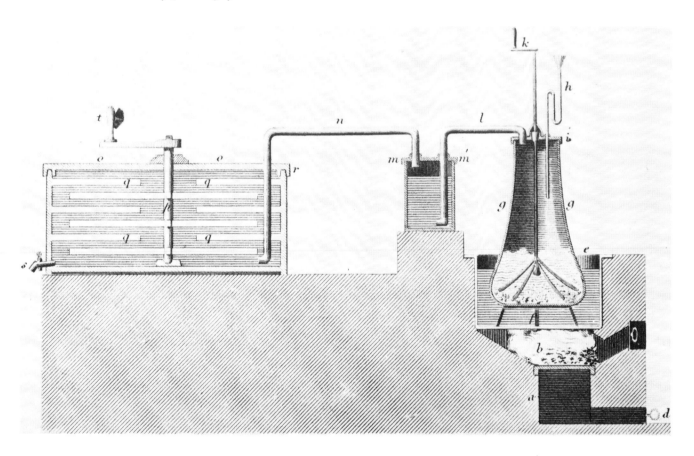

The manufacture of cheap and abundant quantities of bleaching powder suitable for bleaching cloth was another factor which helped to promote the textile boom in the early part of the nineteenth century.

The use of chlorine as a bleaching agent, as recommended by Berthollet (ref 94), had certain practical disadvantages which made the method less acceptable for large scale production as practised in Britain. Chlorine gas in solution weakened the

cloth and had a noxious and unpleasant odour; cases were cited where workmen were poisoned after prolonged exposure to the fumes. The Berthollet process was introduced into Britain by James Watt, who had seen chemical bleaching being carried out in France and instigated a similar method in his father-in-law's bleach-field near Glasgow in 1786. It was a fellow-countryman of Watt, Charles Tennant of Glasgow, who eventually solved the problem of the toxic chlorine fumes;

although Thomas Henry of Manchester worked on similar lines to Tennant, he failed to take out a patent. Tennant's first method was to add lime to the gas in solution, which was then known as oxymuriatic acid; this removed the smell, but did not impair the bleaching properties of the gas. The process was patented in 1798 but the patent, no 2209, was revoked in 1802 on the grounds that the process was not new. In 1799, however, Tennant took out another patent, no 2312, and this was not challenged. The second patent described a method of impregnating slaked lime with chlorine to form a bleaching powder that was inexpensive and harmless to use. Tennant exploited the patent himself in his works at St Rollox, now part of Glasgow. He used sulphuric acid, common salt and manganese oxide to make the chlorine gas, which was brought into contact with the slaked lime in a stone chamber. The constituent materials were cheap and the process was capable of being scaled up for bulk production; so much so that by the 1830s bleaching powder was sold to the textile industry at 3d per lb.

Further Reading

Baines, E., *The History of the Cotton Manufacture in England*, 1835

Chapman, S. D. and Chambers, J. D., *The Beginnings of Industrial Britain*, 1970

ELECTRICITY

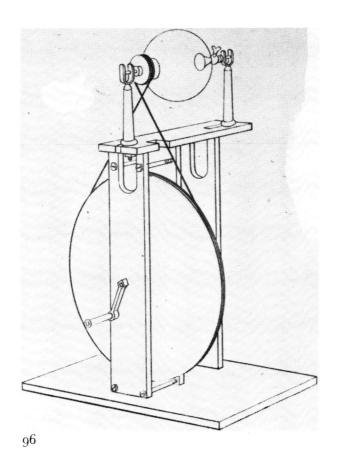

96 ELECTRICAL MACHINE
Francis Hauksbee (d 1713)

Hauksbee was the first to investigate the phenomena of static electricity in the eighteenth century and his friction machine set the pattern for future development.

The electrical friction machine was invented around 1660 by the German physicist Otto von Guericke of Magdeburg. A number of other machines appeared during the eighteenth century which progressively contributed to the understanding of the phenomenon of static electricity, and one of the most significant advances was that made by Francis Hauksbee the Elder in the first decade of the century. Hauksbee held the post of curator of experiments at the Royal Society and was well informed of the work of other European scientists who were investigating various electrical phenomena. He decided to explore further the reasons for the phosphorescent light given off by mercury when shaken in a barometer; a phenomenon which had first been noticed by the Prussian philosopher Bernouilli. Hauksbee had himself invented an efficient air pump and it was not difficult for him to construct the necessary apparatus, which comprised a glass cylinder from which the air could be exhausted at will superimposed on a reservoir of mercury. He noticed that, when a vacuum was maintained in the cylinder, the mercury inside gave off a gentle uniform glow which changed to a series of intermittent flashes when air was admitted.

He continued his experiments with a variety of other substances in place of the mercury and eventually decided to see if similar effects could be obtained by rubbing an exhausted glass vessel in the same way that von Guericke had used a globe of sulphur more than forty years earlier. His apparatus took the form of the electrical machine illustrated. It consisted of a glass sphere from which the air had been exhausted mounted between centres in a form of lathe. An assistant turned the handle to rotate the sphere as rapidly as possible and Hauksbee held his hand on the surface of the glass. A brilliant purple light resulted which was capable of illuminating the room in which the experiment was performed. During the period from 1706 to the end of his life, Hauksbee carried out a series of experiments with his electrical machine, some of which he described in papers presented to the Royal Society and in a pamphlet published in 1709 under the customary ponderous title of *Physico-Mechanical Experiments on various subjects containing an Account of several Surprising Phenomena touching Light and Electricity.* Using two glass spheres he demonstrated examples of electric induction, but the significance of many of these experiments eluded Hauksbee and his contemporaries.

Further Reading
Benjamin, P., *An Intellectual Rise in Electricity*, 1895

97 ELECTRIC LIGHT
Abbé Nollet (1700–70)

Many successful experiments were carried out by eighteenth century scientists to produce electric lights, but more than a hundred years were to elapse before improvements in associated technology made this form of lighting practicable on a commercial scale.

Jean Antoine Nollet was not an ordained priest but he assumed the title of Abbé together with the clerical garb of a minor order of monks in order to obtain security and patronage in the French Court, as well as privileges that would have been beyond the reach of a poor student of science. He became particularly interested in the embryo science of electricity, and constructed an electrical machine which was similar to that of Hauksbee

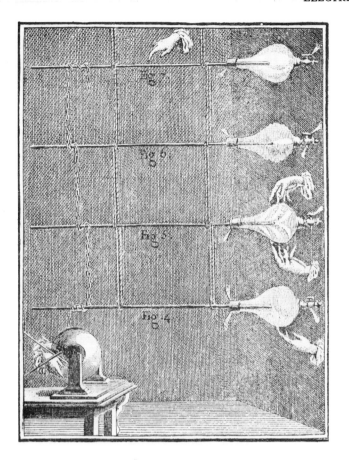

(ref 96), although more ruggedly built. During the years 1740–50 Nollet proceeded to repeat many of the experiments first performed by Hauksbee and other early investigators into electrical phenomena; and in some instances he took the experiment a stage further towards the ultimate practical invention that was to follow in the next century. An outstanding example was Nollet's experiments with electric light. He inserted individual metal conductors inside a series of glass flasks from which the air had been evacuated and linked the other ends of the conductor by a chain to the globe of his electrical machine. When the machine was rotated, the globe itself glowed, as had been the case with Hauksbee's classic experiment, but lights also appeared inside the glass flasks connected by the metal conductors. By placing his hands around the outside surfaces of the flasks Nollet found that he could distort or intensify the luminosity to such an extent that in a darkened room all objects in the vicinity of the apparatus were clearly visible.

Further Reading
Benjamin, P., *The Intellectual Rise in Electricity*, 1895

98 PERMANENT MAGNETS
Godwin Knight (1713–72)

The discovery of a satisfactory way of making permanent magnets led to improvements in the manufacture of the mariner's compass and, eventually, in the nineteenth century to the work of Faraday and his classic experiments on electromagnetic induction.

William Gilbert, the Elizabethan philosopher, is recognised as the true founder of the science of magnetism but, although he was able to make small artificial or permanent magnets, these were only suitable for compass needles. Nearly a century and a half elapsed after Gilbert had published *De Magnete* before a satisfactory method of making larger and more powerful permanent magnets was discovered. The inventor was Dr Godwin Knight, the son of a Lincolnshire clergyman, who after leaving Magdalen College, Oxford, settled in London and is reported to have practised as a physician. He began his researches in magnetism in 1744 and made sufficient progress to deliver a paper on the subject in the following year to the Royal Society. He exhibited a number of powerful bar magnets and presented a further paper which discussed the polar theory of magnetism for which he was awarded the Copley Medal of the society in 1747. Knight devised several methods of making bar magnets and built up a profitable business. He refused to disclose his secrets during his lifetime, and it was only after his death that details were released by his former friends and associates. He appears to have used and improved upon the method of divided touch in which the bar to be magnetised was placed with its ends on the poles of opposite signs of two powerful magnets and two additional magnets were held in position above the bar at an inclined angle. The latter pair with unlike poles in contact were drawn apart and then placed together again, the process being repeated until the bar became magnetic. Another method attributed to Knight was the use of powdered iron oxide which was mixed with linseed oil, moulded into the required shape and baked before being magnetised between the jaws of

a massive compound magnet also made by Knight. The compound magnet, or 'magnetic magazine' as it was called, originally consisted of two halves, each comprising 240 magnets arranged in a rectangular block. It was afterwards re-built in a horse-shoe configuration with N and S poles opposite each other, and it was in this form that Faraday used the apparatus in his famous experiment with the copper disc which was in effect the first dynamo.

Further Reading

Hadfield, D., *Permanent Magnets and Magnetism*, 1962
Taylor, E. G. R., *The Mathematical Practitioners of Hanoverian England, 1714–1840*, Cambridge, 1966

99 LEYDEN JAR
Peter Van Musschenbrock (1692–1761)

The invention of the Leyden Jar, the forerunner of the modern capacitor or condenser, made it possible to store an electrical charge for the first time, and was a major step forward in the development of electrical engineering.

Several experimenters are credited with the invention of the Leyden Jar in the years 1745–6 but it is generally agreed that Peter Van Musschenbrock of the University of Leyden was pre-eminent in this field of research. This is perhaps confirmed by the naming of the apparatus after the location of Musschenbrock's experiments, and this name has been retained ever since. In its original form the jar was partly filled with water and the orifice at the neck was closed with a cork; a wire or nail pierced the cork with the end dipped in the water. The jar was charged by means of a friction machine which was connected to the free end of the wire or nail. The experimenter formed part of the circuit by holding the jar in one hand and touching the wire after it had been disconnected from the friction machine, thus experiencing a violent shock. The Leyden Jar was subsequently used by Franklin (ref 101) and other eighteenth-century experimenters in their investigations, among the most prominent being those of the English scientist Dr William Watson, who established the idea of plus and minus electricity. He also made up circuits of several miles in length and discharged a Leyden Jar through them (ref 100).

Further Reading
Dunsheath, P., *A History of Electrical Engineering,* 1962

100 ELECTRICAL CONDUCTOR
William Watson (1715–87)

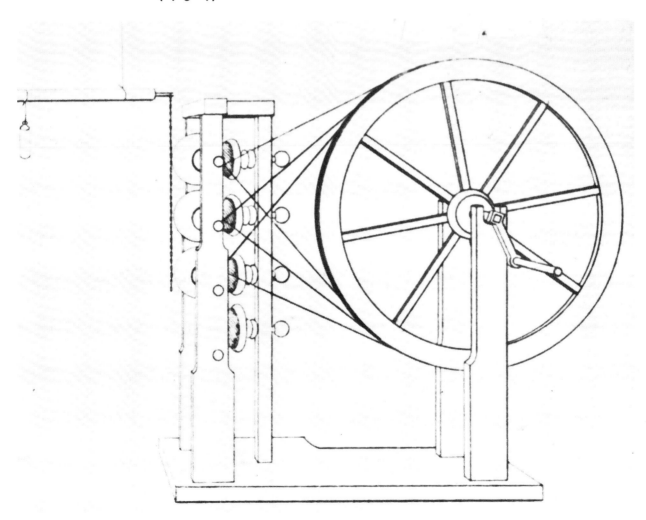

Watson conceived the first electric circuits in which wires instead of chains were used as conductors. His work had far-reaching implications making possible the electric telegraph in the nineteenth century.

Dr William Watson was an apothecary by training who became the acknowledged leader of the English school of electrical experimentalists in the mid-eighteenth century. Watson's work in this field seems to have been stimulated by the invention of the Leyden Jar (ref 99) which provided for the first time means of storing an electrical charge, thus opening up many new avenues of investigation. Like many of his contemporaries Watson constructed an electrical machine which provided a basis for his experiments. In the spring of 1745 he described to the Royal Society the importance of incorporating a metallic conductor in such a machine and using it to concentrate the discharge at a single point. He also ignited alcohol by passing an electric charge through it and fired a musket by the same means. His major contribution to the advancement of electrical knowledge stemmed from an experiment carried out in July 1747. With the help of some friends in the Royal Society he laid a wire along Westminster Bridge—a distance of about 1,200ft— and led each end down to the water's edge. On the Middlesex side a member of the company held one end of the wire while touching the water with an iron rod. Another assistant on the opposite bank held the other end of the wire in one hand and a charged Leyden jar in the other. The circuit was completed by a third man who touched the ball of the jar while holding an iron rod in the water. Immediately the circuit was joined all three intrepid experimenters received a sharp electric shock and a sample of alcohol on one bank was ignited by the current discharged from the Leyden jar on the opposite side. The experiment confirmed the superiority of wire as an electrical conductor, although Watson and his contemporaries did not attempt to differentiate between degrees of conductivity of wire made from various metals. He continued his experiments using circuits of up to four miles in length and at the same time established that a water course, such as a river, was not essential for the return circuit. He was thus the first to appreciate the significance of a 'good earth'. Watson's researches yielded no practical results, but later in the eighteenth century the first rudimentary attempts to send messages through an electric circuit were made by men such as Lesage and Lomond. The Chappe mechanical telegraph (ref 133) brought about a temporary interlude, but ninety years after Watson's initial experiments Cooke and Wheatstone achieved a permanent success and established a new mode of communication.

Further Reading
Benjamin, P., *The Intellectual Rise in Electricity*, 1895

101 LIGHTNING CONDUCTOR
Benjamin Franklin (1706–90)

The lightning conductors suggested by Franklin for the protection of buildings were simple yet effective and during the next two centuries were to save property worth millions of pounds from destruction.

The eighteenth century may have produced more distinguished scientists or more eminent mechanicians than Franklin but he, more than any other, inherited the mantle of the Renaissance man and excelled as a statesman, philosopher, publicist and innovator *par excellence*. He was born in Boston, Massachusetts, and was apprenticed to a printer—his elder brother James—in 1718. After a few years he moved to Philadelphia and then made his first visit to London where he stayed for two years between 1724 and 1726. On his return to Pennsylvania he published the *Pennsylvania Gazette* and his printing activities became so successful that he was able to semi-retire and devote himself to various scientific pursuits. In 1746 Franklin witnessed some electrical experiments in Boston performed by a Dr Spence who had just arrived from Scotland. Some time after he received a glass tube similar to that used in common electrical machines from Peter Collinson in London who was a Fellow of the Royal Society. Franklin proceeded to construct his own electrical machine and with the help of some friends, carried out a number of experiments to investigate the nature of electricity. It was during this time that he carried out his celebrated and highly dangerous experiment of flying a kite in a thunderstorm. This is reported to

have been performed in France at an earlier date, but Franklin was certainly the first to recognise that lightning was an electrical phenomenon. He sent an account of his experiments back to Collinson in London in a series of letters, many of which were read before the Royal Society. Other letters from Franklin to his English acquaintances were first printed in the *Gentlemen's Magazine* and then in April 1751 a collection appeared as a pamphlet entitled *Experiments and Observations on Electricity*. This extended to more than 80 pages and made Franklin famous almost overnight. It was translated into French in the following year, German in 1758 and Italian in 1774, and four English editions were printed before 1769. It was in this pamphlet that Franklin's suggestion of using lightning conductors or rods was first given prominence and in 1769,

when Franklin again visited England, lightning conductors to his specification were fitted on St Paul's Cathedral. These were made of iron and measured 4in by $\frac{1}{2}$in in section. Unfortunately when the cathedral was struck by lightning in 1772 their high electrical resistance caused them to glow at red heat and eventually they were superseded by copper conductors. Controversy also arose over the respective merits of pointed or blunt conductors but most members of the Royal Society agreed with Franklin's view that points were more effective.

Further Reading
An Autobiography of Benjamin Franklin, Watts Edition, New York, 1970
Webster Smith, B., *Sixty Centuries of Copper*, 1965
Dunsheath, P., *History of Electrical Engineering*, 1962

102 ELECTROMETER
William Henley

Henley's electrometer provided a means of measuring electricity quantitively for the first time and was a valuable tool in electrical experiments until superseded by more precise instruments in the nineteenth century.

The early electrical experimenters such as Francis Hauksbee, William Watson and their contemporaries had only the crudest instruments at their disposal, and certainly had no means of measuring the strengths of the electric charges generated by their friction machines. In this respect the electrometer invented by William Henley in 1772 was an important advance, since it not only enabled an electric charge to be measured with a fair degree of accuracy, but it also proved to be a valuable instrument for determining the conductivity of various metals— a programme of research which Henley himself carried out several years before van Marum in Holland (ref 105). The electrometer invented by Henley was simple yet effective. It consisted of a stem of polished boxwood or ivory, an ivory semi-circular scale and a cork ball suspended from the centre of the scale. When the bottom of the stem was electrically charged, the ball was repulsed and moved away so that the magnitude of the charge could be read off from the scale.

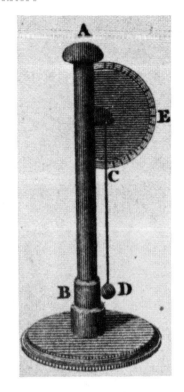

Further Reading
Dibner, B., *Early Electrical Machines*, Norwalk, Conn, 1957

103 ELECTRICAL MACHINE
Edward Nairne (1726–1806)

Edward Nairne's electrical machines included the most powerful built in England during the eighteenth century. Nairne also produced small machines complete with travelling cases for demonstration and therapeutic purposes, as shown in the second illustration.

Those who experimented in the new science of electricity inevitably turned to the mathematical and optical instrument makers to assist in the preparation of their apparatus. Edward Nairne, an instrument maker with a shop at 20, Cornhill, London, was first asked to build an electrical machine by Joseph Priestley, whom he met when attending meetings at the Royal Society. In the years 1771–3 Nairne constructed a machine to Priestley's specifications and then continued to experiment in his own right. He published an account of his first machine in 1773 in a pamphlet entitled *Directions for Using the Electrical Machine as Made and Sold by E. Nairne*; and in the following year he presented a paper to the Royal Society which contained the results of a series of experiments showing the superiority of points over balls as electrical conductors. Nairne continued to build electrical machines and took out patent no1318 in

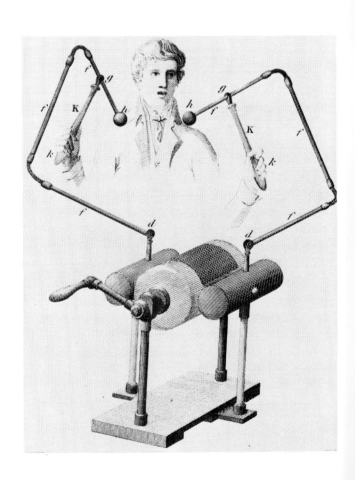

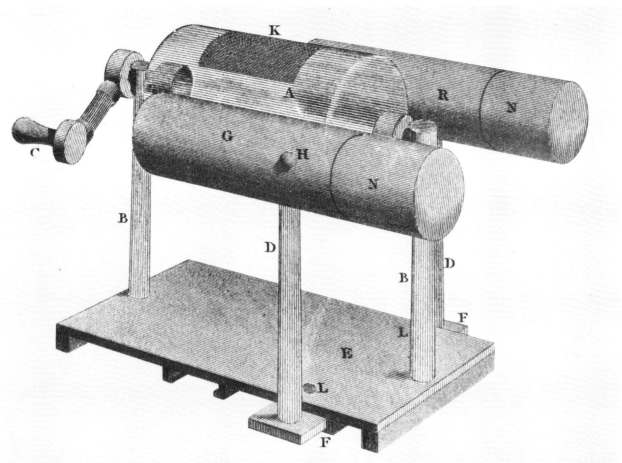

1782 for a machine which he described in the specification as an insulated medical electrical machine. It comprised a central glass cylinder with a friction pad which was flanked by two brass conductors. Leads from the conductors were applied to the patient's body with supposedly beneficial results. In an age when quack remedies were common this attracted considerable attention, and a leaflet describing this apparatus reached its eighth edition by 1796. Nairne also built a very large machine which was only exceeded in power output by that constructed by John Cuthbertson for Van Marum (ref 105). This had a glass cylinder

19in long and 12in diameter with the usual friction pad which pressed on the circumference. The main conductor was a brass cylinder 5ft long with the same diameter as the glass cylinder. A chain connected the bottom of the receiving rod to the pad to complete the circuit. Nairne was able to produce sparks up to 14in long with this machine and, with the aid of a battery of sixty-four Leyden jars, he fused an iron wire measuring 3ft 9in long.

Further Reading
Dibner, B., *Early Electrical Machines*, Norwalk, Conn, 1957

104 ELECTRIC BATTERY
John Cuthbertson (d 1800?)

Cuthbertson's battery consisted of a large number of Leyden Jars. When used in conjunction with Van Marum's electrical machine (ref 105) it provided a unique research tool.

In order that Van Marum and the other experi-

menters associated with the Teyler Museum might have a full range of electrical apparatus at their disposal, John Cuthbertson constructed several large batteries to be used with the massive electrical machine which he had built for the museum. Among these were two of special interest on account

of their size and capacity. The first consisted of twenty-five large Leyden Jars, each 20in high with all inner leads connected with brass rods to a large central sphere. The second battery, which is shown in the illustration with an end view of the electrical machine, comprised nine units each containing fifteen smaller Leyden Jars making a total of 135 jars. It was with this type of equipment that Cuthbertson and Van Troostwijk, another Dutch scientist, succeeded in 1780 in decomposing water into its constituent elements, oxygen and hydrogen, the first known example of electrolysis.

Further Reading
Dibner, B., *Early Electrical Machines*, Norwalk, Mass, 1957

105 ELECTRICAL MACHINE
Martinus van Marum (1750–1837) and
John Cuthbertson (d 1800?)

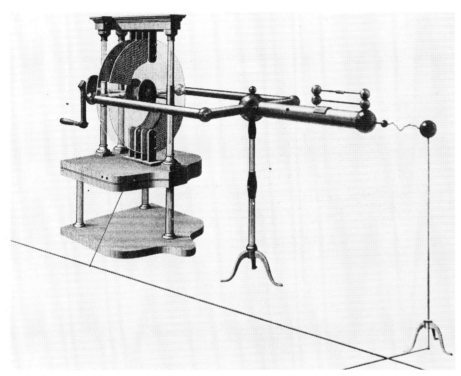

The electrical machine built by Cuthbertson for Van Marum was the largest and most powerful constructed in the eighteenth century. It was used for numerous experiments to establish for the first time the electrical properties of many common materials.

Martinus van Marum was born in Delft and studied medicine and botany at the University of Groningen. Although he practised medicine for a short while, his main interests were in experimental science. He accepted a post at the Teyler Museum in Haarlem, and served the museum as chief experimenter, librarian and director for sixty years. Soon after taking up his duties, he commissioned the English instrument maker John Cuthbertson, who was living in Amsterdam, to build an electrical machine for experimental work. The machine has sometimes been described as Cuthbertson's machine, especially in English text-books, but it probably resulted from close collaboration between the two men. The rotating elements consisted of two glass discs, each 65in in diameter and mounted about 7½in apart on a central shaft. The latter was rotated by a handle with provision for two operators. Rubbing pads of waxed taffeta were fixed, and metal combs were arranged in position about 1½in from the inner surfaces of the glass discs. Glass insulators, which must have made the apparatus even more fragile, were also fitted, and a series of cylindrical brass conductors carried the charge to the contact points. Van Murum was able to make sparks about 2ft long with the machine; after many trials the most effective solution for treating the rubbing pads was found to be an amalgam of mercury, tin, zinc and gold which must have added considerably to the cost of the experimental programme. The equipment, however, was put to good use; Van Marum conducted many experiments in which wires made from various materials were fused, and from the data obtained he was able to conclude that copper was the best conductor of the cheaper metals and should be used for lightning conductors. Franklin (ref 101) had not considered this aspect; indeed many early lightning conductors were made of wrought iron and were reported to have glowed red hot when a building was struck by lightning.

Further Reading
Dibner, B., *Early Electrical Machines*, Norwalk, Conn, 1957

106 TORSION BALANCE
Charles Coulomb (1736-1806)

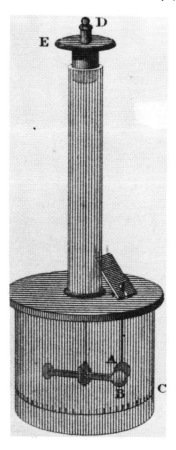

Coulomb's torsion balance was a major advance in electrical instrumentation and led to the discovery of the inverse square law which states that the repellent force between two electrically charged spheres is inversely proportional to the square of the distance between their centres—the basic law of electrostatics.

During the last two decades of the eighteenth century, a number of significant advances were made in electrical instrumentation which were eventually to lead to the voltaic pile (ref 107) and the new field of electro-chemistry. One of the most brilliant inventions of this era was the sensitive torsion balance of Charles Coulomb, a former French army engineer who settled in Paris, where he investigated various problems relating to the properties of materials. His torsion balance, which followed from his studies of torsional rigidity, was constructed in about 1784. The instrument consisted of a glass cylinder with a graduated scale around the lower part of the circumference. The top of the cylinder was covered by a round glass disc with two holes, one at the centre and the other at a pre-determined distance from the centre. A second slender cylinder was placed over the centre hole and a torsion micrometer was arranged at the top. From this was suspended a fine silver wire which carried at the lower end a horizontal reed reinforced with sealing wax. The reed had a pith ball attached to one end and a small paper disc was arranged as a counter-weight at the other end. A second pith ball was lowered through the other hole in the disc and positioned close to the first ball. When both balls were similarly charged the force of the subsequent repulsion could be measured by the angle of the displacement twist or torque, and by using extremely fine wire Coulomb was able to make the instrument sensitive to a force of only 0.0005 dyne. In 1785 he described his invention in a paper presented to the Academy of Sciences and in his conclusions he proposed the inverse square law which supported his experimental results. Coulomb's work in this field was confirmed by Joseph Priestley and Henry Cavendish in England.

Further Reading
McCloy, S. T., *French Inventions of the Eighteenth Century*, Kentucky, 1952
Dibner, B., *Early Electrical Machines*, Norwalk, Conn, 1957

107 ELECTRICAL BATTERY
Alessandro Volta (1745-1827)

The voltaic pile or battery, which could produce a steady electric current, ushered in a new era in the development of electrical engineering and thereafter interest in the phenomenon of static electricity as produced by friction machines gradually decreased.

Alessandro Volta, who was born at Como in northern Italy, carried out his most important electrical research while employed as a professor at the University of Pavia. In 1775 he invented the electrophorus by which a small charge of static

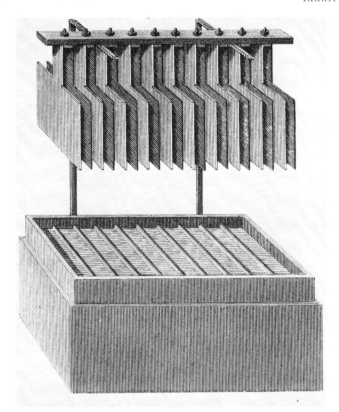

electricity could be increased many times through the manipulation of a mechanical apparatus. He continued to work on a wide variety of electrical experiments and for a time was influenced by the work of Galvani at the rival University of Bologna. Galvani, as a result of his celebrated series of experiments on frogs, had postulated the theory of 'animal electricity' which implied that electricity could be excited by living organs. Volta, however,

soon turned his attention to studying the effects of dissimilar metals when forming part of a circuit; he thereby rediscovered the phenomenon observed twenty-five years earlier by Sulzer that dissimilar metals in contact when placed on the tongue produced a curious tingling sensation. His experiments showed that a wide area of metal such as a ·flat plate gave the best results and that if he added a second pair of plates separated by a fibrous diaphragm the electric current, as he termed the discharge, was increased. Subsequently he added additional pairs of plates until he had constructed the first voltaic pile. Details of this historic apparatus were published in Italy in 1799 and a few months afterwards Volta, who was a Fellow of the Royal Society, announced his discovery in a formal letter to Sir Joseph Banks, the society's president. Volta apparently experimented with various combinations of metals but concluded that silver and zinc with a solution of any neutral or alkaline salt gave the best results. As an insulator he recommended a piece of pasteboard or leather. Although his apparatus may have been crude it contained all the basic elements of the modern battery and by increasing the number of plates up to sixty a powerful discharge sufficient, to quote Volta's own words, 'to give shocks as high as the shoulder' could be produced.

Further Reading
Dunsheath, P., *A History of Electrical Engineering*, 1962

PRINTING AND PAPERMAKING

D. JUNII

JUVENALIS

ET

AULI

PERSII FLACCI

SATYRAE.

BIRMINGHAMIAE:
Typis JOHANNIS BASKERVILLE.
MDCCLXI.

108 TYPE DESIGN
John Baskerville (1707–75)

The Baskerville type-face combined an elegance and sense of proportion with a practical adaptability which made it suitable for almost every sort of printed matter. After a long period of neglect, it is even more popular today than it was in the eighteenth century.

John Baskerville was born in Worcestershire but moved to Birmingham, where he spent most of his working life. He taught calligraphy for four years as a young man, and this led him to an interest in typography and eventually printing. Like other master printers of his day, he was intimately concerned with every aspect of his trade, and he made notable contributions to the general advancement of printing through the preparation of improved inks and the use of paper smoothed by being pressed through hot copper plates. It was in type design however that he made his greatest contribution. His type-face, which bears his name, was particularly graceful, and it set a new standard for eighteenth-century typography, not only in Britain but on the continent and in North America, where Benjamin Franklin was a confirmed admirer. The first book bearing Baskerville's imprint was a quarto *Virgil* which appeared in 1757. This was followed by some fifty other books and miscellaneous works which were all outstanding in their excellence of presentation. Baskerville, indeed, was the first to insist that typography took precedence over the work of the illustrator and engraver, and the easy readability of his type has long been recognised. His views were not generally accepted by many of his contemporaries, and towards the end of his life his business declined. He tried unsuccessfully to sell his entire equipment to a continental press, but the prospective purchaser was unable to meet his price of £8,000. After his death his widow sold his punches, matrices and presses for 150,000 francs to Pierre-Augustin Beaumarchais, the publisher of Voltaire's works. Baskerville types eventually found their way into many French printing works, although during the last twenty years certain of the original types have been returned to England as the result of the generosity of a leading French printing firm. The Baskerville type-face has regained its popularity in the present century and is today highly regarded by printers and publishers.

Further Reading
Steinberg, S. H., *Five Hundred Years of Printing*, 1959
Cummings, A. D., 'Printing and Papermaking Machinery', *Engineering Heritage*, Vol 2, 1966

109 PAPER-MAKING MACHINE
Francois Louis Robert (1761–1828)

Until the end of the eighteenth century almost all paper in Europe and America was made from linen and cotton by a complex process requiring a considerable amount of labour. Robert's papermaking machine was a major advance since wood pulp was the stock material and paper could be made to a fixed width and unlimited length. Its effect on the expansion of learning in the following century was immeasurable.

Dissatisfaction with the established methods of papermaking was shown in many quarters during the eighteenth century, particularly as the demand for paper grew through the proliferation of newspapers, books and pamphlets of all types. In 1732 René de Réaumur proposed that paper should be made from wood, citing as evidence the paper-like material made from wood fibre by certain species of wasp. Réaumur never carried out any experiments to test his theory, and it was left to a German pastor, Jacob Schaffer, to prove conclusively that it was possible to manufacture paper from the fibres of a variety of trees and plants. It was not until the closing years of the century, however, that a practical method of making paper from wood pulp was evolved. This depended on the ingenious machine invented by François Robert which, in improved and enlarged form, is still universally used today. Robert was in turn a soldier, printer and schoolteacher, but it was while employed as a printer by the celebrated publishing house of Didot that he first began his attempts to build a papermaking machine. His prime consideration was said to be the construction of a machine that would require as few operatives as possible, an

objective which met with the encouragement and approval of his employer. Robert's first attempt in 1797 was a failure, but in the following year he produced another machine for which he applied for a patent on 9 September 1798. After a thorough examination by the Conservatory of Arts and Trades, who exercised jurisdiction over inventions during the First Republic, Robert was granted a patent in 1799 for the customary period of fifteen years and was awarded 3,000 francs. The key component of the machine was a continuous belt of copper or brass wire mesh which revolved over a horizontal frame fitted with a series of metal rollers that were cranked by hand. Wood pulp or paste was spread over the wire belt which was then given an oscillating motion to shake off surplus water from the pulp into vats below. The pulp then passed through two felt rollers which removed some remaining moisture and was then dried in a separate compartment. Paper up to 12ft wide and 50ft long could be made in this way. It is apparent that

Robert's machine, as patented in France, was far from perfect, and neither Robert nor his employer, Didot Saint-Léger, were able to exploit it successfully. A British patent no3568 was obtained by Didot and the machine was improved in Britain by John Gamble, who was a brother-in-law of Didot, Bryan Donkin and the Fourdrinier brothers, Henry and Sealy. The brothers purchased the patent rights in 1804 and, although in time they lost control of the machine, it is still known by their name. The unfortunate Robert attempted to manufacture machines on a commercial basis in France but, after losing a great deal of his own money and that of his supporters, he retired to the small country town of Dreux, where he spent the last fifteen years of his life as an impoverished schoolmaster.

Further Reading

McCloy, S. T., *French Inventions of the Eighteenth Century*, Kentucky, 1952

126

110 LITHOGRAPHY
Alois Senefelder (1771-1834)

The invention of lithography occurred in the closing years of the eighteenth century, but it is only in the last fifty years that the true significance of this new concept in printing has been fully appreciated.

Alois Senefelder was born in Prague in 1771 and at his father's insistence became a law student before turning to the more precarious profession of acting. He also wrote plays in his spare time and it was his attempts to publish his own plays that brought him into contact with the printing industry. He studied letterpress printing and stereotyping, as well as etching on copper plates, and his attention was drawn to the merits of limestone as a printing surface. This in itself was not a new innovation in

Bavaria, where Senefelder was living, as a form of relief printing in which the stone was etched with nitric acid had been practised for several years. Senefelder applied himself to improving the process; he discovered a method of giving the stone a higher polish and formulated an ink composed of wax, soap and lampblack which could be easily wiped away from the surface once the image had been covered. It was, in fact, the ink which paved the way for the modern lithographic technique. In collaboration with a publisher in Munich named Steiner, Senefelder began to put his new process to some practical use and among other things printed a number of music sheets and an illustrated prayer book. He continued to experiment and it was in

this period between the years 1796–98 that he discovered the basic principles of lithography, namely the affinity and rejection of certain liquids when brought into contact. He noticed that a viscous liquid, such as a solution of gum, prevented the ink from attaching itself to the stone and, when he drew an image on the stone with soap and moistened the surface with gum water, the ink only adhered to those areas covered with the soap. Senefelder subsequently went into partnership with Johann André, a music publisher, who paid him two thousand florins for a description of the process and the right to exploit it. This agreement was apparently only a local one for Senefelder came to London in the summer of 1800 and on 20 June 1801 a British patent, no2518, was taken out in his name.

Further Reading
Twyman, M., *Lithography (1800–1850)*, 1970

111 PRINTING PRESS
Charles, third Earl of Stanhope (1753–1816)

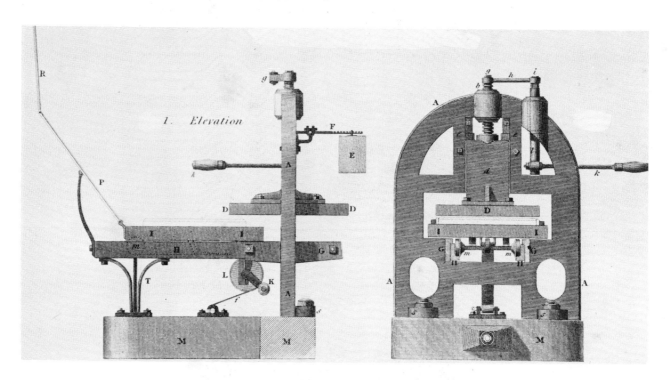

The Stanhope or iron press made it possible to print a large forme in one pull instead of the two that were formerly required.

Charles Stanhope was endowed with tremendous energy and an appetite for knowledge which enabled him to pursue successfully the seemingly incompatible careers of politician, philosopher and scientist. He was aware of his limitations as a practical engineer, and for his experiments relating to the improvement of the printing press he engaged the services of Robert Walker, a mechanic of Vine Street, Piccadilly. Stanhope began his work in the last decade of the eighteenth century but his press was not completed in 1804. The principal improvements were the adoption of cast iron in place of wood and the use of a combined screw and lever movement. The bed of the machine was increased in size, and it became possible to print a form in a single operation where two would have been required in the past; this was due to the greatly increased pressure that could be exerted as a result of the use of iron. Stanhope presses were used for printing *The Times* in the early years of the nineteenth century, but were superseded by the steam-driven König press in 1814.

Further Reading
Steinberg, S. H., *Five Hundred Years of Printing*, 1959

POTTERY
AND GLASS

112 PORCELAIN MANUFACTURE
William Cookworthy (1705–80)

High quality porcelain was manufactured in China for centuries before the secret was known in the West. First Böttger in Germany in 1709 and later William Cookworthy in Plymouth laid the foundations of the European porcelain industry in the eighteenth century.

Credit for being the first to manufacture porcelain successfully in England is usually given to William Cookworthy of Plymouth, although there is evidence to show that porcelain was made at Bow several years before Cookworthy began his experiments although the practice was discontinued. Cookworthy, who was born at Kingsbridge in Devon, started a druggist business, but retired at the age of forty and became a travelling Quaker minister. It was during one of his visits to Cornwall in the late 1740s that he discovered china clay or kaolin deposits and also deposits of soapstone on Tregonnin Hill, near Helston. It was known that these were the principal ingredients of porcelain and the kaolin was already being used for refractory furnace bricks in the neighbouring tin smelters. Cookworthy started to experiment in the manufacture of porcelain but was unable to produce ware of satisfactory quality for a number of years. Many of his difficulties were due to the use of coal in the kilns which smoked and stained the porcelain. Eventually he turned to a wood-fired brown stoneware kiln and found fresh supplies of clay and stone at Carlogass near St Austell which were of higher quality than those originally worked near Helston. It was not until 1768 that he obtained a patent for the exclusive manufacture of porcelain, and in the same year he established a factory at Plymouth. His works soon employed between fifty and sixty people, and a skilled painter and enameller was introduced from Sèvres. There was a good demand for porcelain, and quantities were even exported to America. After two years, however, the manufactory was moved to Bristol where the pottery tradition was stronger, and in 1774 Cookworthy transferred the full patent rights to Richard Champion, who had previously played an important part in the running of the business. Champion petitioned parliament in 1775 to have the patent extended and, although he was partly successful, the enterprise did not prosper. Champion was declared bankrupt in 1778 and the patent right was purchased by a company in Staffordshire who thereafter had access to the supplies of Cornish clay and stone.

Further Reading
Watney, B., *English Blue and White Porcelain of the Eighteenth Century*, 1965

113 PYROMETER
Josiah Wedgwood (1730–95)

Considerable progress was made during the eighteenth century in the measurement of temperatures up to the boiling point of water but little attention was given to the problem of assessing furnace temperatures. Wedgwood's pyrometer—crude as it may have been—enabled the potter to regulate his oven and exercise a degree of control over the firing operation.

Josiah Wedgwood, more than any other man, was responsible for the dominating position gained by the Staffordshire pottery industry during the second half of the eighteenth century. Wedgwood was not only a master craftsman and an innovator unsurpassed by his contemporaries, he was also a man of great business acumen. Despite an increasing preoccupation with commercial affairs as his firm expanded, he remained closely involved with the

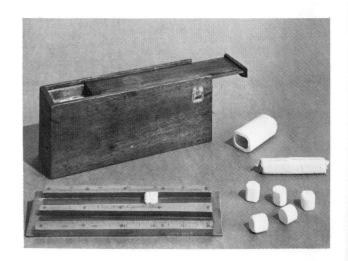

practical problems of pottery manufacture all his life. Wedgwood appears to have conducted experiments to check furnace temperatures in the years 1780–1. He published some of his results in a paper delivered to the Royal Society in 1782 and in January of the following year he was elected a Fellow of the Society. In 1784 and 1786 he published two further papers on the subject which together summarised his work. His methods were crude but effective, his apparatus consisting merely of a series of clay cones which were placed in the furnace and then measured for shrinkage on removal. After some practice an experienced craftsman could assess whether the temperature was right for firing by checking the degree of shrinkage. Although its limitations were obvious, the method remained in use in some potteries for over a hundred years.

Further Reading

Carter, E. F., *Dictionary of Inventions and Discoveries*, 1966

Shaw, S., *History of the Staffordshire Potteries*, 1829, reprinted Newton Abbot, 1971

114 OPTICAL GLASS
Peter Louis Guinard

Guinard's glass making technique resulted in greater availability of high quality glass for lens manufacture.

Considerable progress was made in the development of optical instruments during the eighteenth century, but their manufacture on a large scale was hindered by difficulties in producing glass of homogeneous optical quality. Eventually this was overcome in the closing years of the century by an inventive Swiss bell founder, Peter Louis Guinard. His technique, first introduced in 1798, was relatively simple, but it revolutionised glass founding. He suggested that the molten glass should initially be stirred in the crucible and then allowed to cool in the conventional manner. The contents of the pot were broken down and the most suitable lumps of glass were then re-melted and the cycle repeated. The molten glass was finally poured into suitable moulds, cooled and subsequently ground and polished into lenses of the desired curvature. One disadvantage of the process was that the yield from one particular pot could not be accurately forecast but, as glass manufacturers acquired more experience, the amount of inferior glass discarded after the first melt decreased, and up to 1,000lb of high quality glass could be produced in one operation.

Further Reading

Venis, E., 'Glass Manufacture', *Engineering Heritage*, Vol 2, 1966

AERONAUTICS

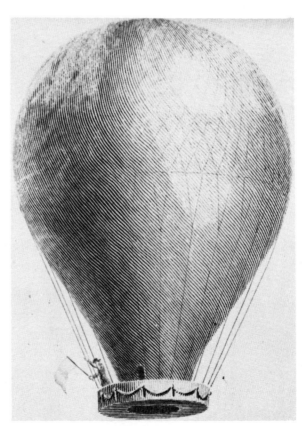

115 HOT AIR BALLOON
Joseph Montgolfier (1740–1810) and
Etienne Montgolfier (1745–99)

The first successful contrived flight of any living creature and the first manned flight were achieved by the Montgolfier brothers with their hot-air balloons in 1783, thus heralding the air age and a new dimension for the intrepid to explore.

Joseph Montgolfier, whose family business was paper making, first experimented with a hot air balloon in November 1782. His earliest model had an envelope made of silk and was so small that it could be demonstrated inside a room. On being filled with hot air, it rose to the ceiling—a feat which so impressed Joseph's brother Etienne that he became a partner in the enterprise. Together the brothers made three more experimental balloons, the largest having a diameter of about 35ft. It was at this stage that they began experimenting with a combination of cloth and paper as the balloon fabric, a material that they were to use successfully with their full scale balloons in due course. On 25 April 1783 the largest of their experimental

balloons rose 1,000ft and then descended gently to earth about ¾ mile from the starting point. This dispelled any lingering doubts about the feasibility of flight and a full size balloon was then made with an envelope capacity of 22,000cu ft. Again the trial was a success, the balloon attaining a height of 6,000ft and covering a distance of 1½ miles in a flight of ten minutes duration. The brothers were then ready to demonstrate their invention publicly and a site was selected at Versailles. Yet another even larger balloon was made with an envelope capacity of 37,500 cu ft. The occasion became something of a Roman holiday with a large crowd assembling in front of the Royal Palace to witness the spectacle. A wicker basket was attached to the balloon in which were placed a sheep, a cock and a duck, and the final preparations were punctuated with booming cannon shots. On being released the balloon rose to 1,700ft and drifted two miles before landing; a contemporary engraving shows the envelope tilted at rather an alarming angle but the occupants of the basket were none the worse for wear when they landed. The Montgolfiers now had a formidable rival in the field in the person of the physicist, Jacques Charles (ref 116) and they worked feverishly to attain their final objective—the first manned flight. Success was theirs on 20 November when a balloon of their design and manufacture with a capacity of 79,000 cu ft made a flight of about five miles from Paris with two occupants, a scientist Pilâtre de Rozier and an aristocrat le Marquis d'Arlandes. Montgolfier balloons vied for popularity with hydrogen balloons for many years and, although the latter gradually gained ascendancy, the hot-air type continued to be used because of its lower operating costs.

Further Reading
Rolt, L. T. C., *The Aeronauts*, 1965

116 HYDROGEN BALLOON
Jacques Charles (1746–1823)

The hydrogen balloon of Jacques Charles was not only the true forerunner of the military balloons and the dirigibles of the nineteenth and twentieth

centuries, but it led directly to the less spectacular scientific balloons which continue to provide valuable meteorological data.

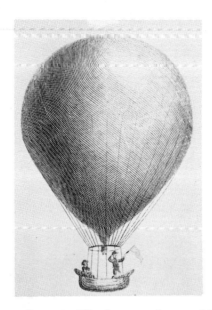

Jacques Charles was thwarted by the Montgolfier brothers (ref 115) in the race for the honour of making the first unmanned and manned balloon flights, but his balloons were technically superior to those of his rivals. As a physicist and member of the Academy of Sciences his approach was inevitably more scientific, and he not only used the recently-discovered gas hydrogen to inflate his balloon but adopted rubberised silk for the fabric. He first used a small experimental balloon which, after some difficulty, he filled with hydrogen generated by the action of sulphuric acid on iron filings. The process was slow and costly, but Charles was rewarded with success at his first attempt in the summer of 1783. The balloon rose swiftly to about 3,000ft and was carried 15 miles from its starting place in the Champs de Mars in Paris. He immediately began to construct a second and much larger balloon in order to make a manned flight, but the honour of being first went to the Montgolfiers on 20 November 1783. Charles was undeterred, and with a companion

took off from the Tuileries Gardens before a vast crowd on 1 December 1783. The conditions were ideal and the balloon made a flight of two hours' duration, landing 27 miles from Paris. Charles' second balloon incorporated several technical improvements which were subsequently adopted in later balloons and air-ships; these included a relief valve to permit gas to escape at high altitudes and another manually controlled valve fitted in the north pole position which could be opened inwards when the aeronaut wished to descend. A hydrogen balloon was used by the Italian Lunardi to make the first ascent in England on 15 September 1784, but initially the hot air balloon was the more popular type. As supplies of hydrogen became more readily available, the gas balloon gained ascendancy over its rival and was almost universally adopted for military and scientific purposes.

Further Reading
Rolt, L. T. C., *The Aeronauts*, 1965

117 AIRSHIP
Jean Baptiste Meusnier (1754–93)

Only the lack of a suitable prime mover prevented Meusnier from building a successful dirigible balloon, and many of his ideas, particularly the balloonet, were adopted by his successors in the second half of the nineteenth century.

Jean Baptiste Meusnier was one of several brilliant French military engineers who made significant contributions to the development of technology

outside their chosen field. When still a young man he perfected a method of distilling sea water and independently of Argand (ref 124) invented a circular lamp wick. He is, however, chiefly remembered for his design of a dirigible balloon which can be considered as the first practical step in the evolution of controlled flight. Meusnier outlined his ideas in a paper presented to the French

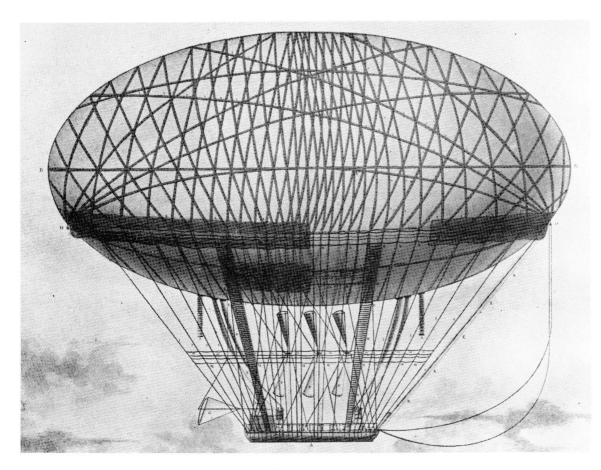

Academy on 3 December 1783—only two days after Charles' historic flight in a hydrogen balloon from the Champs de Mars. It was in this paper, entitled *Mémoire sur l'équilibre des Machines Aéro-statiques*, that Meusnier with unerring preconception made a number of fundamental proposals that were afterwards adopted as standard practice for the design of non-rigid and semi-rigid dirigibles in the second half of the nineteenth century. Perhaps his most significant contribution was that of the balloonet which was essentially a separate gas bag carried inside the balloon. In his design the inner bag was to be inflated with hydrogen and the space between the inner and outer bags filled with air. This was intended to conserve gas and provide a means of regulating the height. It also helped to preserve the elliptical shape of the envelope which Meusnier had concluded produced the least resistance to motion through the air. To carry the occupants Meusnier proposed to use a boat-shaped car slung underneath the envelope with three propellers arranged in line midway between the car and the envelope. This was another important development,

since most other early dirigible designers who followed Meusnier, including Cayley, favoured paddles and other impractical solutions. In 1784 Meusnier produced a design for a dirigible in ellipsoidal form that was 260ft long and had a capacity of 60,000cu ft. There is no record that it was ever built, but a smaller dirigible with a capacity of 30,000cu ft constructed according to Meusnier's ideas made a flight from St Cloud on 15 July 1785. The crew comprised the Duc de Chartres and the brothers Robert who subsequently gained fame as balloonists. According to reports the primitive rudder provided was fairly ineffective, and there was little semblance of control. Meusnier was killed in action against the Prussians at Mayence in 1793, and nearly sixty years were to pass before Giffard made his successful powered flight in a steam-driven dirigible.

Further Reading
Hildebrandt, A., *Airships Past and Present*, 1910
Rolt, L. T. C., *The Aeronauts*, 1965

118 THE AEROPLANE
Sir George Cayley (1773-1857)

Many leading air historians agree that Cayley was the basic originator of the modern aeroplane. Although most of his work was accomplished in the nineteenth century, he began his researches in 1796 and before the turn of the century he produced a drawing of a fixed wing aeroplane which initiated a chain of development that culminated over a hundred years later in the successful machines of the Wright Brothers.

Sir George Cayley, a Yorkshire baronet, lived and carried out most of his research work at Brompton Hall near Scarborough. His first experiments with heavier-than-air flight were conducted in 1796 and concerned a model helicopter. Two Frenchmen, Lannoy and Bienvenu, had previously made a similar model which consisted of two twin-bladed rotors operated by a bowstring mechanism. Cayley's version was even simpler, four feathers being used for each rotor in place of silk, with the central boss formed by a cork. A bow-string mechanism again produced a contra-rotating motion that was sufficient to sustain a short flight. Cayley's next project is generally known as his 'silver disc' since the only record of its existence is in the form of an engraved silver disc now in the Science Museum in London. On one side is a diagram showing the forces of drag, lift and thrust and on the reverse is a sketch, bearing Cayley's initials, of a small fixed-wing glider. The salient features of the design are the clearly defined wing unit with the fuselage below in which a seat was provided for the pilot and the tail unit comprising horizontal and vertical control surfaces. Forward motion was to be obtained by a pair of manually-operated flappers in place of a airscrew. In the same year, 1799, Cayley produced a drawing for a full-size version of the machine complete with dimensions and weights. There is no evidence to show that the 'silver disc' plane was ever built but many of its features were incorporated in later Cayley models and gliders that flew successfully.

Further Reading
Gibbs-Smith, C. H., *Sir George Cayley's Aeronautics (1796–1855)*, Science Museum Handbook, 1962
Gibbs-Smith, C. H., *Aviation—An Historical Survey*, 1970

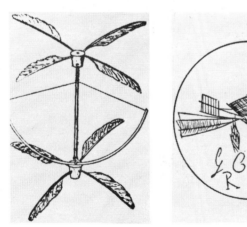

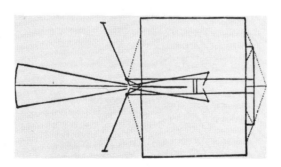

ORDNANCE

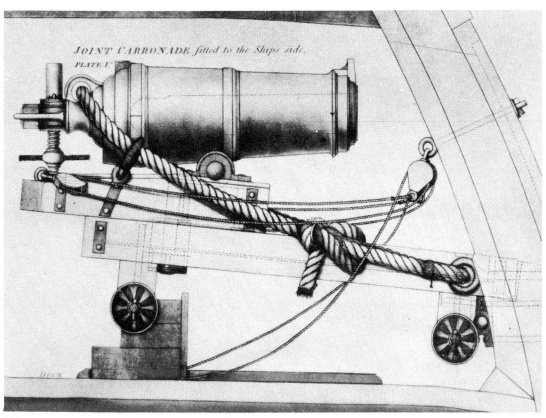

JOINT CARRONADE fitted to the Ships side.
PLATE I

DECK

119 CARRONADE
Robert Melville (1723–1809) and
Charles Gascoigne

The short large calibre carronades possessed tremendous destructive power when used against wooden ships o' the line; they made a major contribution to the establishment of British Naval supremacy during the Napoleonic Wars.

Although there is no doubt about the place of origin of this formidable piece of ordnance, a mystery still exists regarding the original inventor. The carronade was made by the Carron Company of Falkirk between 1778 and 1852. It was a most destructive weapon being nick-named the 'smasher' and was highly suited for short range combat as dictated by the naval tactics of the period. In the naval version the barrel was mounted on a wooden carriage running on four small wooden wheels (not as shown in the photograph). The breech end was supported on a wooden wedge and a vent was provided to carry a train of powder from the touch-hole at the upper part of the breech. Two thick ropes were attached to either side to restrict the recoil when the gun was fired. The carronade was rapidly adopted for arming merchant ships and by the Admiralty for the Royal Navy. The Admiralty was quick to exploit the advantages of this relatively small, compact gun which could fire a heavier shot than cannon previously in service and which was considered to be infinitely more manoeuvrable. Fewer men were required to work it

and—most important of all—its rate of fire was higher than the standard naval guns of the period. One report claims that a 68 pounder carronade could be fired three times as fast as an earlier 32 pounder. The invention of the carronade has been attributed to General Robert Melville, a former infantry officer, and also to Charles Gascoigne, partner and manager of the Carron Company. Melville's claims are not substantiated by contemporary writers although he is known to have had close contacts with the company and attended the

120 SHRAPNEL SHELL
Henry Shrapnel (1761–1842)

The spherical case or shrapnel shell, as it came to be called, made an important contribution to British victories in the Peninsular War and at Waterloo—a fact that was readily acknowledged by Wellington and his staff.

Henry Shrapnel was commissioned as a lieutenant in the Royal Artillery in July 1779. After service in Newfoundland he returned home and spent the remainder of his service career engaged on problems of ordnance. In 1784 he began a series of experiments at his own expense to produce a hollow iron sphere or shell containing a number of bullets, together with a bursting charge and a fuse, to cause the shell to burst at a pre-determined point in the trajectory. Spherical case, as it was initially known, was intended to produce the same devastating effect at long range as ordinary case shot did at up to three hundred yards. In the first design the bursting charge was in direct contact with the bullets, and premature explosions were sometimes caused by friction. The bursting charge was not added until the shell was at the gun position to avoid the risk of an explosion in the ammunition wagon, and this inevitably slowed down the rate of fire. Modifications were subsequently made and the charge was enclosed in a tube, thus eliminating the early difficulties. Shrapnel spent a considerable amount of time and money on the project, but finally had the satisfaction of having his shell adopted by the Army in 1803–4, on the recommendation of the Board of Ordnance. It was an unqualified success, first in the attack on Surinam in 1804 and a few years later in the Peninsular War. Wellington himself wrote to Shrapnel after Vimiera in August 1808 to tell him how his shell had contributed to

first trials. Undoubtedly the task of supervising the casting and boring of the new gun rested with Gascoigne and much of the credit for its success is due to him. Patrick Miller, the patron of William Symington, was also associated with the project but his assistance was confined to financial support and acting as a broker between the company and the Admiralty.

Further Reading
Campbell, R. H., *The Carron Company*, Edinburgh, 1961

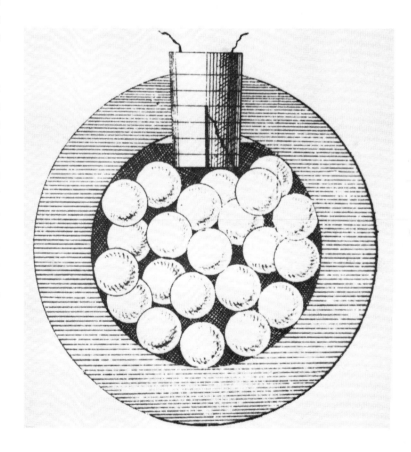

the defeat of the enemy, and it was also used effectively at Waterloo. Shrapnel retired from the army in 1825 with the rank of Major-General and was awarded a pension of £1,200 a year for his invention.

Further Reading
Hughes, B. P., *British Smooth-Bore Artillery*, 1969

DOMESTIC

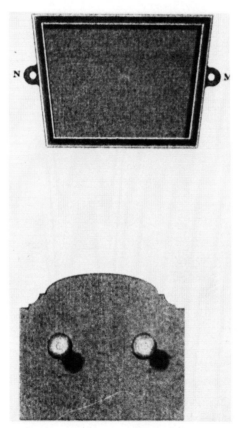

121 STOVE FOR ROOM HEATING
Benjamin Franklin (1706–90)

Franklin's stove was an assembly of iron castings and could be manufactured cheaply and in appreciable quantities. It became popular on both sides of the Atlantic, especially in New England where it helped to fight the privations of the severe winters.

Although Benjamin Franklin is chiefly celebrated for his fundamental researches into the phenomena of static electricity (ref 101), his inventive mind ranged over a whole multitude of problems which had a bearing on contemporary life in eighteenth-century America. The severe winters suffered by Pennsylvania and the New England colonies caused much hardship and discomfort, and stimulated his invention of a cast iron stove or fireplace in 1742. Franklin made a model which he presented to a Mr Robert Grace who was an iron founder as well as a personal friend. The stove was an assembly of relatively simple iron castings which Grace was able to produce in some quantity without difficulty. Its principal advantages were economy in fuel and a greater thermal efficiency; an air box was provided in which cold air was drawn in and heated by the flue gases before being discharged into the room. Franklin published a pamphlet extolling the virtues of his stove and Governor Thomas of Pennsylvania offered to allow him the sole vending rights for an unspecified period of years. Franklin declined for reasons which, he stated in his autobiography, resulted from his desire to be of service to his fellow men. Certainly other manufacturers were not slow to take advantage of his philanthropy and his stove was sold in large numbers throughout the Colonies. An ironmonger in London patented a similar stove after reading Franklin's pamphlet and was reported to have made a sizeable fortune on his investment. Possibly because of Franklin's fame in other avenues of human endeavour, the stove was still known as the Franklin stove in England, even in the nineteenth century.

Further Reading
The Autobiography of Benjamin Franklin, Watts edition, New York, 1970

122 SHEFFIELD PLATE
Thomas Bolsover (1704–88)

The practice of covering a base metal such as copper with a thin layer of silver enabled manufacturers of silverware to reduce prices and expand sales.

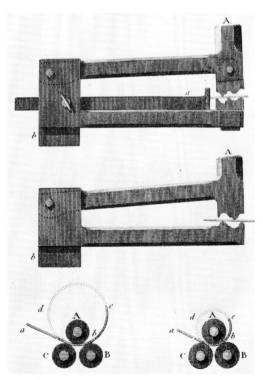

The invention of Sheffield Plate was one of those fortuitous accidents of history which are exploited by men of foresight and business acumen. In 1742 Thomas Bolsover, a Sheffield silversmith, was repairing a knife handle made of silver and copper when he accidentally overheated it and caused the metals to flow so that the silver formed a coating over the copper. He realised the commercial value of his discovery and began to make and sell a variety of small objects which, although largely made of copper, had the appearance and quality of silver. Bolsover thus laid the foundations of the Sheffield plate industry, which lasted in Britain for about a hundred years, until it was superseded by electroplating. Others besides Bolsover subsequently contributed to the craft of plate manufacture.

Joseph Hancock found that by finishing the two plates separately and heating them under pressure they could be made to adhere, and in 1784 George Cadman introduced the practice of soldering a silver bead along the edge of the sheets to ensure that the copper core was completely hidden. The composite material was ductile and easily worked, as is evident from the many beautiful examples of the craft that have survived to the present day. All the typical metalworking techniques of the time were used and some of the common tools can be seen above.

Further Reading
Smith, B. Webster, *Sixty Centuries of Copper*, 1965

123 LEVER LOCK
Robert Barron

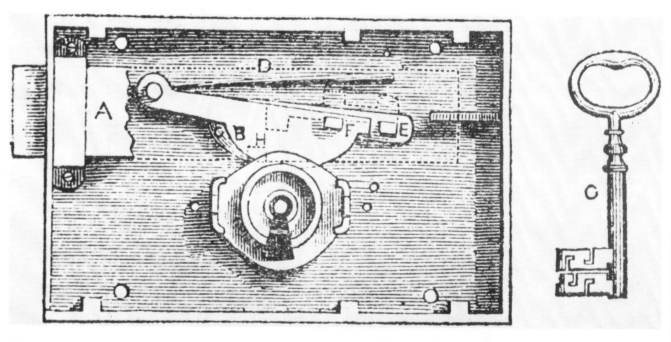

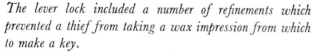

The lever lock included a number of refinements which prevented a thief from taking a wax impression from which to make a key.

During the last quarter of the eighteenth century, the Society of Arts through its Mechanicks Committee did much to encourage and improve the standard of lock manufacture, a policy no doubt induced by the activities of numerous thieves and picklocks who dwelt in every town and city. In 1778 Robert Barron of Hoxton was granted patent no 1200

for a lever lock which greatly reduced the chances of the picklock since it virtually nullified the established technique of taking a wax impression and using it to make a key. In Barron's lock, movement of the bolt was prevented by a series of lugs which projected into grooves cut in the side of the bolt. The lugs, which were located on the sides of the pivoted lever, had to be raised to a certain height to permit the bolt to move, and this could only be achieved with the correct key. The ingenious feature of the design was that those parts of the levers which made contact with the key were not accessible from the keyhole. The thief could only ascertain the correct distance to which the levers had to be raised by trial and error. Originally only two levers were fitted, but this number must have proved insufficient because later locks had up to six levers. Picking a lock, even for an experienced thief, could then be a lengthy business, and the lever lock did much to reduce the incidence of theft.

Further Reading

McNeil, I., *Joseph Bramah, A Century of Invention*, Newton Abbot, 1968

124 OIL LAMP
Aimé Argand (1755–1803)

Argand's lamp, because of its circular wick and glass chimney, gave a much brighter flame and was a great improvement over other lamps used during the latter part of the eighteenth century.

Aimé Argand was the son of a prosperous Swiss watchmaker. He inherited his father's manual dexterity and after completing his education at the University of Geneva he moved to Paris, where he quickly became associated with Lavoisier, Fourcroy and other prominent members of the Academy of Sciences. While still in his early twenties, he invented a greatly improved method of wine distillation for which he was awarded a prize of 120,000 livres. He moved temporarily to the Lower Languedoc district and it was during his time there that he conceived the idea of a new form of wick for oil lamps based on the technique used for heating spirits of wine. In 1780 he made his first model which incorporated the circular wick but had a metal chimney. Argand took his lamp back to Paris and tried to get it adopted for lighting the Paris streets. For some unknown reason he did not apply for a patent from the Academy of Sciences, and when little interest was shown in his invention he crossed the Channel to England and obtained a British patent no1425. He returned to Paris early in 1784 and completed his invention by substituting a glass chimney. This part of the lamp was subsequently protected by an Order of the Council of State of 30 August 1785, and Argand was given the exclusive right to manufacture and sell glass chimneys for lamps for a period of fifteen years. This was not sufficient to stop others from pirating his design, particularly the circular wick, and Argand gained very little financially from his invention. He left France after the Revolution and eventually died in poverty in Geneva. One of the disadvantages of the original Argand lamp was that

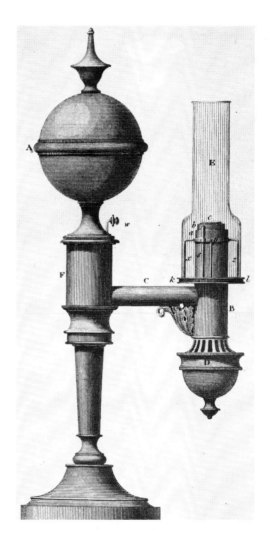

the spherical oil reservoir, from which the oil was fed to the wick through a horizontal tube, created a shadow when the lamp was lit. This was afterwards overcome by introducing a circular reservoir placed around the wick.

Further Reading

McCloy, S. T., *French Inventions of the Eighteenth Century*, Kentucky, 1952

125 LOCK
Joseph Bramah (1748–1814)

Joseph Bramah's lock was far in advance of anything available at the time and was virtually impossible to pick. However a high degree of precision was required in its manufacture which made it costly to produce until Bramah and his foreman Henry Maudslay designed a series of special tools to accomplish the more intricate operations.

The Bramah lock is a typical example of an essentially eighteenth-century solution to a mechanical problem. It was intricate and ingenious, as one might expect from the agile brain of a master inventor-craftsman, but it did not lend itself to mass production with the tools available at the time. Bramah first became interested in locks when he started to attend meetings of the Mechanicks Committee at the Society of Arts towards the end of 1783. As a member Bramah was able to examine a number of locks submitted by many of the foremost locksmiths of the day—a trade that was flourishing due to the activities of the numerous gangs of thieves to be found in every town and city. It was generally agreed that the most efficient lock was that of Robert Barron of Hoxton (ref 123) which had been patented in 1778. Bramah soon appreciated that even the Barron lever lock had one or two weaknesses which made it vulnerable to the attentions of an experienced thief. These stemmed from the fact that to release the bolt all the levers were raised a constant distance. Bramah's solution, as outlined in his patent no1430 granted to him in 1784, was to use the same lever principle but ensure that each lever had to be raised by a different amount from that of its neighbour before the bolt could be released. This required a complex mechanism of notched levers or sliders which were arranged in the form of a barrel to give greater compactness. The sliders were placed radially on the barrel and the latter had to be rotated before the bolt could be released. Production of this lock on a commercial scale soon ran into difficulties, and about six years elapsed after the patent was granted before Bramah locks became available. This was after Henry Maudslay had joined his employment, and it is generally agreed that it was Maudslay who solved many of the production problems by designing and manufacturing a series of special machine tools,

some of which have survived and are included in the Science Museum collection. Eventually lock manufacture became one of Bramah's principal activities at his Pimlico works in London. In 1798 a second patent was granted to Bramah which virtually extended his original 1784 patent a further fourteen years.

Further Reading
Rolt, L. T. C., *Tools for the Job*, 1965
McNeil, I., *Joseph Bramah A Century of Invention*, Newton Abbot, 1968

126 GAS LIGHTING
William Murdock (1754–1839)

Others before Murdock experimented with gas illumination during the eighteenth century, but Murdock alone enjoyed the support of an engineering company such as Boulton, Watt & Co. It was his work which led directly to the formation of the Gas, Light and Coke Company in the early years of the next century and the widespread use of gas for illumination and eventually heating.

In 1791 William Murdock, who was serving as erector and maintenance engineer in Cornwall for Boulton, Watt & Co., first began experiments to determine the feasibility of gas illumination. He had been preceded by others in Britain and on the continent, including George Dixon of Cockfield, Co Durham in 1760, Minckelers in Holland in 1784 and, only a few years previously, by Lord Dundonald who in 1787 had illuminated Culross Abbey with coal gas flares. Murdock, who worked in a workshop attached to his house in Redruth, considered a number of gas feedstocks including wood and peat

before he decided that coal was the most economic and that it also provided the best light. He designed and built a pilot generating, storage and distribution plant together with fixed and portable burners, and in the next year, 1792 he used coal gas to light the rooms in his house and the offices attached. The retort was sited in the yard and the main inlet pipe was led through a hole in a window frame to the ceiling of the principal room; other pipes led from this point to burners in various parts of the house, the furthest being 70ft from the cast iron retort. Murdock carried out further experiments at Neath Abbey Ironworks between November 1795 and February 1796, but when he communicated the results of his work to Boulton, Watt & Co, the company was reluctant to take out a patent in his name because of the unexpired patent held by Lord Dundonald. This stalemate lasted several years and Murdock returned to his native Scotland for a time. In 1798 he was back in Soho on the permanent staff and Watt himself began to take an interest in his experiments. It so happened that the company had been engaged in the manufacture of retorts and purifiers for a certain Dr Beddoes of Bristol who was investigating the curative properties of oxygen and hydrogen when

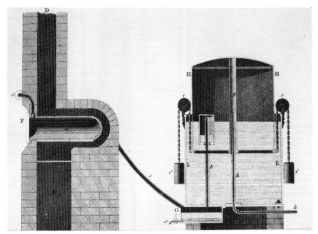

inhaled. It was therefore comparitively simple to redesign this equipment and incorporate Murdock's own ideas evolved in Redruth. News of Lebon's work on similar lines in France at last spurred the company to produce the apparatus on a commercial basis, and the first demonstration of its effectiveness was the illumination of the Soho factory to celebrate the short-lived Peace of Amiens in March 1802. Murdock was awarded the Rumford Gold Medal by the Royal Society in 1808. He was not concerned with the formation of the Gas, Light and Coke Company in 1812 but retained a link through Samuel Clegg, the inventive and resourceful chief engineer, who had served under him as an apprentice at Soho and who was responsible for many of the basic developments in the early years of gas engineering.

Further Reading

Smiles, Samuel, *Lives of the Engineers: IV Boulton and Watt*, 1861

Braunholtz, W. T. K., 'Gas Engineering', *Engineering Heritage*, Vol 2, 1966

MISCELLANEOUS

Fig. 7.

127 FIRE ENGINE
Richard Newsham (d 1743)

Richard Newsham's fire engine was the first to produce a constant jet of water rather than a succession of short squirts. This type of fire appliance remained in service for many years until finally superseded by the steam fire engine of the early nineteenth century.

Memories of the disastrous Great Fire of London haunted the City merchants for decades afterwards, but it was not until Newsham's fire engine appeared that an appliance was available which provided any measure of protection. Newsham himself was originally a pearl button manufacturer in the City of London and had first-hand knowledge of the problems which faced the City Fathers. He took out his first fire engine patent, no439, in 1721 and this was followed by a second patent, no479, in 1725. His appliance consisted essentially of a double-cylinder reciprocating pump and a large air vessel mounted on a cart that was sufficiently narrow to pass through a standard size doorway. The pumps were actuated by a pivoted beam worked by groups of men at each side, with the pistons moving up alternately. The air vessel produced a constant jet of water which was reported to be powerful enough to break a window and, after one of the earliest trials, the *Daily Journal* of 7 April 1726 reported that a stream of water had been thrown as high as the grasshopper on the Royal Exchange, which was 160ft from the ground. Newsham supplied a number of engines to insurance companies and also to the chief provincial towns. After his death in 1743 the business was carried on under the name of Newsham & Ragg, engine-makers of Cloth Fair, and a number of fire engines were supplied to the Royal Navy. The Newsham engine was effective when dealing with small fires but was unable to cope with major conflagrations in a closely built-up city area where

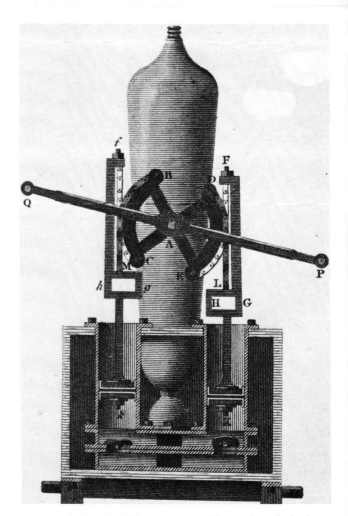

many of the buildings were still made of timber. This is evident from the chronicle of disastrous fires that continued right through the eighteenth century in London, in which three or four hundred houses could be destroyed as a result of one fire.

Further Reading
Jackson, W. E., *London's Fire Brigades*, 1966
McNeil, I., *Joseph Bramah, A Century of Invention*, Newton Abbot, 1968

128 VENTILATION OF SHIPS AND BUILDINGS
Stephen Hales (1677–1761)

Hales' system of ventilation had immediate and beneficial effects whenever it was used, especially in the overcrowded prisons where the death-rate was significantly reduced.

Stephen Hales was another eighteenth-century

ecclesiastic who was concerned equally with the practical improvement of the living conditions of his fellow men as he was with their spiritual welfare. Most of his inventions and improvements related to matters of hygiene and agriculture. He was himself

148

form until about 1740 when a ventilating system with a forced draught provided by bellows was installed in the ship *Sorbay*. Prior to Hales' innovation the only undertakings that were ventilated to any degree were mine workings, where practical necessity demanded that the foul air should be displaced by fresh as the shafts were driven progressively deeper. Hales was successful in getting his system installed in other ships and in granaries and prisons. He was even able to persuade the French to fit ventilators in prisons where English prisoners of war were incarcerated, as well as in the Newgate and Savoy prisons in London. The ideas put forward by Hales stimulated others, and Martin Triewald in Sweden and the French highway engineer Pommier also constructed similar systems. Hales himself continued to strive for improvements and in his seventy-fifth year applied a small windmill to the task of ventilating houses. A later version, which is illustrated, shows that rotary motion of the vanes was converted into reciprocating motion by means of a crank which in turn may have been used to actuate a form of bellows. Hales published *A Description of Ventilation* and *A Treatise on Ventilation* to promote his ideas in 1743 and 1758.

a brilliant botanist and was elected a Fellow of the Royal Society in 1718. He made a number of fundamental discoveries in botany, and was awarded the Copley medal by the society in 1739. His most celebrated invention did not appear in practical

Further Reading
Dictionary of National Biography, Vol **XXIV**, 1890

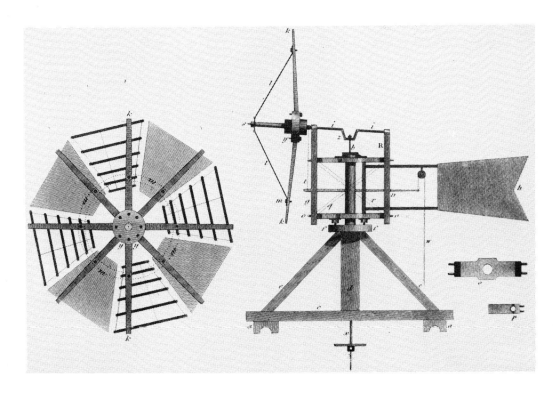

129 STEEL MILL
Carlisle Spedding (d 1755)

The steel mill yielded a shower of sparks that were sufficient to illuminate mine workings where candles could not be used. It was nevertheless inherently dangerous and was discarded when safety lamps became available in the early nineteenth century.

Carlisle Spedding was one of the most celebrated colliery managers in the north of England during the first half of the eighteenth century. He held the post of manager at the Whitehaven mines owned by the Lowther family in Cumberland for thirty-seven years, during which time he sank pit shafts to a greater depth than had been previously attempted in Britain and also extended the mine workings under the sea bed. He installed no less than four Newcomen engines at Whitehaven in order to keep the workings free of water and made strenuous efforts to improve the ventilation in the mines under his charge. During the years 1740–50 he invented the steel mill or Spedding mill, as it came to be known, in order to provide a degree of illumination in sections of a mine where candles could not be used because of fire damp. The invention was in fact based on a grave misconception, since the sparks given off by this mill were sufficient to cause an explosion under certain conditions in unventilated mines. The mill consisted of a thin steel disc which was rotated by a handle at great speed and brought into contact with a piece of flint. A continuous shower of brilliant sparks was emitted which gave sufficient illumination for miners to work at the coal face. The mill was

usually operated by a child and was worn strapped round the neck and waist. Carlisle Spedding was eventually killed in a fire damp explosion and his son James, who succeeded him as manager of the Whitehaven collieries, introduced additional improvements in underground ventilation in an effort to eliminate further diasters. The Spedding mill continued to be used, however, until it was superseded by the safety lamps of Davy and George Stephenson in the early years of the next century.

Further Reading
Galloway, R., *A History of Coal Mining in Great Britain*, 1882, reprinted Newton Abbot, 1969

130 CALCULATING MACHINE
Charles, third Earl of Stanhope (1753–1816)

Stanhope's calculating machine was an important step in the progression which began with Pascal in the seventeenth century and finally culminated in the modern computer.

Few inventors in the eighteenth century, except perhaps Smeaton and Bramah, were more versatile than the third Earl of Stanhope, and yet his accomplishments are generally overlooked in most surveys of the period. Stanhope enjoyed all the privileges that were the inherent right of an English nobleman. He was educated at Eton and travelled widely on the continent as a young man. He showed his precociousness when he was only eighteen by writing a paper in French on the pendulum, for which he was awarded a prize by the Swedish Academy. He became involved in politics at an early age and for a time was a close friend of William Pitt, whose sister became his first wife. Despite the fairly exacting demands of his political life, he managed to pursue his interests in science and engineering, and among his earliest inventions were two calculating machines which were made in the years 1775–7. Stanhope employed a skilled mechanic, James Bullock, to build the machines to his specifica-

tions. The first was of the push-pull type with an upper frame that could slide sideways in a guide frame which itself could reciprocate backwards and forwards. The mechanism was intricate yet compact, comprising trains of gears, wheels, a series of toothed setting wheels and a number of racks arranged axially around a cylinder. Dials were provided at the front and the machine could be used to multiply and divide. The second machine completed in 1777 was a modification of the first and both included certain features which were incorporated into later calculating machines in the nineteenth century. The general verdict on both machines is that they were efficient to operate but were slow to set up, and this reason, plus the fact that they must

have been highly expensive, probably accounted for the fact that only two were built. Both models were owned some years later by Charles Babbage and it seems reasonable to assume that they influenced his own work. Stanhope, whose life has been described as one of continuous industry and toil, turned his attentions elsewhere and made notable contributions in such diverse fields as optics, cement manufacture, canal building, electricity and printing (see ref 111).

Further Reading

Last, J., 'Digital Calculating Machines', *Engineering Heritage*, Vol 2, 1966

131 COPYING MACHINE
James Watt (1736–1819)

Watt's machine for copying letters became a standard item of office equipment until superseded late in the nineteenth century by the combination of the typewriter and carbon paper.

As the business interests of Boulton, Watt & Co expanded so did the correspondence between the

partners, one of whom was often away from the Soho Works. Watt was aware of the advantages of keeping copies of his letters to Boulton and also records of the company's business dealings with mine owners and other customers for the pumping engines in various parts of the country. The practice of making hand-written copies was irksome to him,

151

and accordingly he set about devising a copying machine. He was an accomplished chemist, and two of his friends, Joseph Black and Joseph Priestley, were among the greatest chemists of the day. He experimented with ink and sheets of unsized paper to see whether acceptable copies could be produced by merely pressing a sheet on top of a letter immediately after it had been written. After about two months in 1779 he found a satisfactory formula and sent the results of his experiments to Boulton for his approval. He then designed two types of copying press, one actuated by a screw and the other by a roller. In February 1780 he took out patent no1244 to cover both presses and the special ink. A separate company, James Watt & Co, was formed to manufacture the presses, and James Keir, who had previously established a glass works at Stourbridge and was a fellow member of the Lunar Society, joined the enterprise as a partner. Only the roller press was produced at first, but it proved to be an immediate success. Although the impression on the copy paper was in reverse, the paper itself was so thin that the document could be read from the other side. Boulton displayed his customary business acumen and took one of the first presses to London, where it was demonstrated before members of

parliament, bankers and City merchants. It won general approval, although the bankers had some misgivings, fearing that it would facilitate forgeries. Eventually this prejudice was overcome, and in the first year of business the new company sold a hundred and fifty machines. The screw model came into production at a later date, and both versions, with modifications, remained in demand until the last quarter of the nineteenth century.

Further Reading
Rolt, L. T. C., *James Watt*, 1962

132 ROLLER BEARINGS
John Garnett

Roller bearings were initially preferred to ball bearings because they were easier and cheaper to make. Garnett's patent more or less summarised the developments made during the earlier years of the eighteenth century.

There is evidence that roller bearings were used fairly widely during the eighteenth century, particularly in mill work. Sully, the celebrated horologist, is reported to have made a ship's chronometer with roller bearings, and in 1734 patent no543 was granted to Jacob Rowe for a device utilising friction wheels with axles bearing on other wheels. The definitive patent relating to roller bearings is, however, generally regarded as that taken out by John Garnett—no1580—in 1787. Garnett's bearing incorporated caged rollers located by spring retaining rings with an outer race to ensure axial alignment. It thus contained most of the fundamental features of modern roller bearings. A subsequent development at the end of the century was the invention

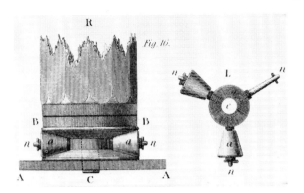

of the roller thrust bearing with conical rollers, as shown in the second illustration. In this particular example taken from a mill, only three rollers were fitted, so that the load on each must have been considerable.

Further Reading
Naylor, H., 'Bearings and Lubrication', *Engineering Heritage*, Vol 2, 1966

133 THE TELEGRAPH
Claude Chappe (1763–1805)

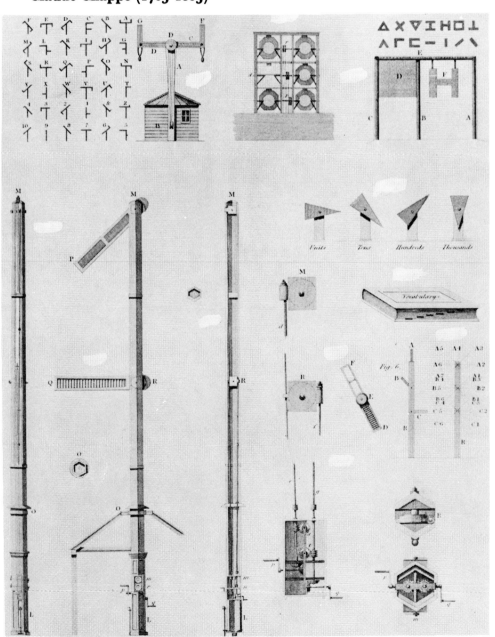

Chappe's visual telegraph was rapidly developed in France and other countries of Western Europe because of its obvious military potential. Trained operators could transmit messages at remarkably high speeds and the system remained in service well into the nineteenth century until superseded by the electric telegraph.

Although a number of inventors attempted with varying degrees of success to transmit messages by means of electricity during the eighteenth century, the only practical telegraphic system was that devised by Claude Chappe, using an essentially mechanical or visual method. Chappe himself initially attempted to use electricity but it was not until he resorted to a visual technique in 1790 that he made any progress. With the aid of his three brothers, Chappe, who was a native of Brûlon near Le Mans, constructed two stations about a quarter of a mile apart with indicators the arms of which could be arranged in a variety of positions. The brothers found that with practice they could send messages to each other quite quickly by means of a simple code. A public demonstration was held on a larger scale in March 1791, and later that year the Chappe brothers went to Paris where they set up an experimental station in the area now known as the Place de l'Etoile on the Champs Elysées. Revolutionary Paris gave little encouragement to inventors and their station was speedily torn down one night by the mob. Chappe, who by this time had spent 40,000 livres of his own money, finally managed to bring his invention to the attention of the National Convention where it was favourably received; and a decision was taken, after further trials, to build a chain of sixteen stations to connect Paris with the army of the north at Lille. It was undoubtedly the obvious military advantages of rapid communication between the capital and the armies in the field that persuaded the rulers of the Republic to exploit Chappe's invention. A second line was completed to Strasbourg in 1798, and in the same year the line to Lille was extended north to Dunkirk. Each line comprised a series of stone towers. Above each tower was a heavy pole about 10ft high. A crossbar approximately 14ft long, pivoting at the centre, was attached to the top of the pole, and arms, each about 6ft long, were fixed to the ends of the crossbar. Wires linked the arms to the operator's control position. Chappe formulated a code of 9,999 words, each represented by a number which could be indicated according to the position of the arms. At night four lanterns were placed at the ends of the signal arms, and it was only during foggy weather that the system was inoperative. Visual telegraphic systems were invented independently in Sweden and Britain during the closing years of the eighteenth century, but the Chappe system is generally considered to have been the most efficient; messages from Lille for example being transmitted 150 miles in two minutes.

Further Reading
McCloy, S. T., *French Inventions of the Eighteenth Century*, Kentucky, 1952

134 BALL BEARINGS
Philip Vaughan

Philip Vaughan's patent for a radial ball bearing had a significant effect on the development of many forms of wheeled transport in the nineteenth century, but it could not be properly exploited until machine tools were available for grinding the balls accurately and cheaply.

A type of ball race was used in Classical times, and many of the Renaissance inventor craftsmen were familiar with the principle of rolling contact surfaces. Benvenuto Cellini is known to have used wooden balls to support his statue of Jupiter. Plain journal bearings were satisfactory for the slow moving steam engines of the eighteenth century, and little interest was shown in the ball race until towards the end of the century, one of the earliest known examples being a large thrust bearing, 2ft 10in in diameter with cast iron balls $2\frac{1}{2}$in diameter, fitted by an unknown millwright to a post mill near Norwich in 1780. Fourteen years later patent no2006 was granted to Philip Vaughan

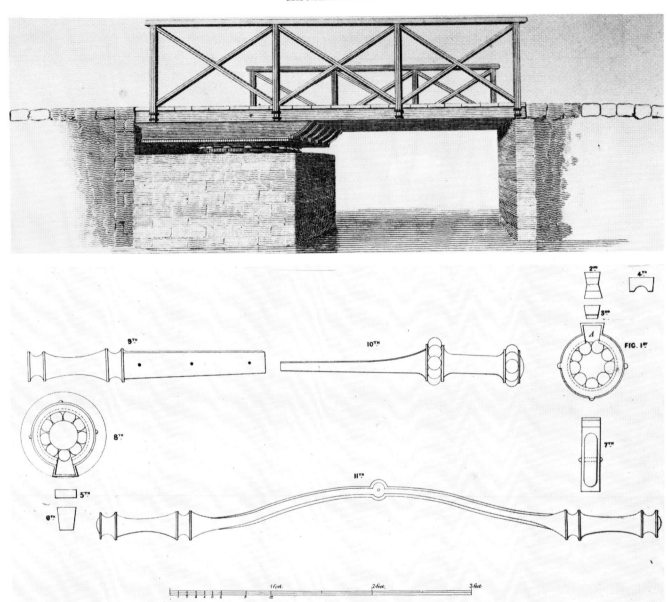

for a radial ball bearing to be used in conjunction with a vehicle axle. The balls were uncaged, and rolled directly in the stationary raceways formed in the shaft. The revolving outer races were provided with a filling slot through which the balls could be inserted, and which was closed with a wedge-shaped insert. Vaughan's specification was too exacting for the machine tools of the period, and widespread adoption of this type of bearing did not take place until the 1860s, when it began to be widely used

by bicycle manufacturers. In the late eighteenth and early nineteenth centuries, the applications were principally confined to large thrust bearings such as the type illustrated in the picture of the swing bridge.

Further Reading
Naylor, H., 'Bearings and Lubrication', *Engineering Heritage*, Vol 2, 1966

135 ROPEMAKING MACHINE
Joseph Huddart (1741–1816)

Huddart's machine produced ropes of superior strength and durability, and large quantities of rope from his factory

in Limehouse were supplied to HM Dockyards for the Fleet during the Napoleonic Wars

Joseph Huddart was a captain of an East Indiaman and spent the first part of his life at sea in service of the East India Company. He had first-hand experience of the frequent breakages of ships' cables and found that this was a result of most of the strain being taken by the outer yarns in the strands. When he retired he decided to design a machine which would rectify this defect. His main objectives were to keep the yarns separate and draw them from bobbins revolving on pins, in order to maintain the twist while the strand was formed. The yarns were passed through a register plate and a tube was used for compressing the strands. Studdart took out a patent no 9512 in April 1793 but nearly seven years elapsed, during which time he made a series of experiments at Maryport in Cumberland, before he eventually offered his invention to the East India Company. The company referred the matter to the private contractors whom they employed to supply ropes. These companies were reluctant to change their traditional methods, and the Admiralty showed a similar inertia when Huddart offered his machine to them. Fortunately Huddart found a group of friends who were willing to support him financially, and he set up a factory at Limehouse to manufacture ropes on his own account. In August 1799 Huddart was granted a second patent, no2339, which regulated the angle of twist given to a rope and introduced modifications that enabled ropemaking to be carried out in a confined space. These developments were incorporated into the machinery installed at Limehouse which soon received orders to supply ropes to HM Dockyards and a number of collieries. Huddart ropes were immediately recognised for their superior performance over traditionally manufactured ropes and instances were cited where they outlasted three ordinary ropes of comparable size. A Huddart machine installed in Deptford Dockyard remained in service until 1855, and when his patents expired his machine was generally adopted throughout the ropemaking trade.

Further Reading
Tyson, W., *Rope*, 1966

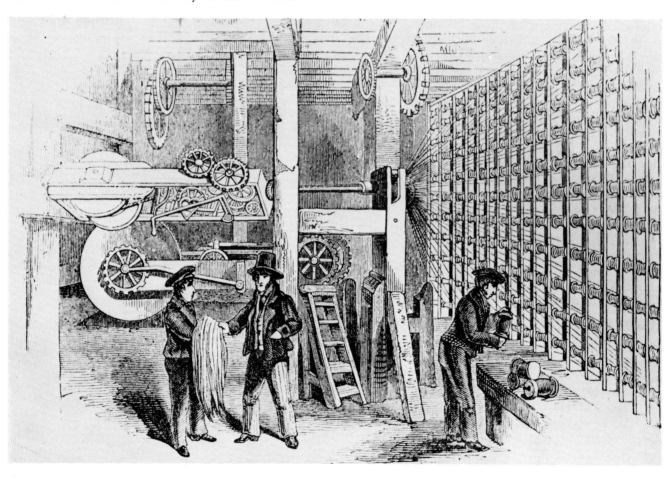

ACKNOWLEDGEMENTS

G. E. Fussell, *The Farmers Tools*, 1, 4, 5

Radio Times Hulton Picture Library, 2, 8, 38, 45, 46, 101, 128, 135

The Science Museum, South Kensington, 15, 17, 21, 22, 24, 34, 63, 70, 74, 77, 82, 84, 85, 86, 87, 88, 89, 91, 99, 109, 113, 114, 117, 126, 130, 131, 132, 134

E. Baines, *The History of Cotton Manufacture in England*, 37, 40

Conservatoire des Arts et Métiers, Paris, 39, 71

R. Buchanan, *Practical Essays on Mill Work*, 75

J. P. M. Pannell, *Illustrated History of Civil Engineering*, 80

S. D. Chapman and J. D. Chambers, *The Beginnings of Industrial Britain*, 92, 94

Park Benjamin, *The Intellectual Rise in Electricity*, 97

S. H. Steinberg, *Five Hundred Years of Printing*, 108

The Victoria and Albert Museum, South Kensington, 112

Charles H. Gibbs-Smith, *Aviation*, 118

Carron Company Ltd, 119

Maj-Gen B. P. Hughes, *British Smooth Bore Artillery*, 120

Ian McNeil, *Joseph Bramah*, 123

University of Sheffield, 129

All other illustrations are from the plate volumes of Rees's *Cyclopaedia*.

INDEX